Avian Influenza and Newcastle Disease

Ilaria Capua • Dennis J. Alexander
Editors

Avian Influenza and Newcastle Disease

A Field and Laboratory Manual

Foreword by Joseph Domenech and Bernard Vallat

Springer

Editors

Ilaria Capua
Head, Virology Department
Director, OIE/FAO and
National Reference Laboratory
for Newcastle Disease and Avian Influenza
Istituto Zooprofilattico Sperimentale delle Venezie
Legnaro, Padua, Italy
icapua@izsvenezie.it

Dennis J. Alexander
Former Director EU
OIE/FAO Reference Laboratory
for Avian Influenza and
Newcastle Disease
Veterinary Laboratory Agencies
Weybridge, UK
d.j.alexander@vla.defra.gsi.gov.uk

The Editors and the Publishers wish to thank Papi Editore for the permission to re-use part of the figures published in the volume *A colour Atlas and Text on Avian Influenza (I. Capua, F. Mutinelli)*. © 2001, Papi Editore

Cover illustration: courtesy of Amelio Meini

ISBN 978-88-470-0825-0 Springer Milan Berlin Heidelberg New York
e-ISBN 978-88-470-0826-7

Library of Congress Control Number: 2008937917

Springer is a part of Springer Science+Business Media
springer.com

Cover design: Simona Colombo, Milan, Italy
Typesetting: C & G di Cerri e Galassi, Cremona, Italy
Printer: Printer Trento Srl, Trento

Printed in Italy
Springer-Verlag Italia, Via Decembrio 28, I-20137 Milan, Italy

This atlas is dedicated to the memory of Dr. Giovanni Vincenzi, former Head of the Veterinary Services of the Veneto Region, Italy, as a heartfelt acknowledgement of his invaluable contribution to the field of veterinary public health.

Foreword

The farming of poultry is one of the primary means of supplying human beings with high quality protein. As a consequence, over the last century or so, there has been a shift in industrialised and in several developing countries from predominantly rural farming to intensive large-scale poultry farming.

Viral diseases are very common in poultry but, due to the non-pathognomonic characteristics of their signs and lesions, they are frequently misdiagnosed. Infections of poultry with highly pathogenic avian influenza (HPAI) virus or Newcastle disease (ND) virus are generally accompanied by high mortality and severe economic losses for the poultry industry, not only from the loss of animals as a direct result of disease but also from trade restrictions and embargoes that may be imposed. In addition, some strains of HPAI have implications for human health. ND and HPAI are therefore considered the two most important diseases of poultry.

The Food and Agriculture Organisation of the United Nations (FAO) and the World Organisation for Animal Health (OIE) have always recognised the critical nature of these diseases and have thus responded to epidemics in a proactive manner, issuing guidelines and recommendations, organising missions to countries where disease occurs to assess and evaluate situations and providing expertise and support to member countries.

In recognition of the challenge faced by the veterinary community to improve the animal health status worldwide, this manual has been produced as an instrument to support laboratories as well as official and private veterinary services in the diagnosis and management of outbreaks of avian influenza and ND. An improved diagnostic effort carried out at a global level will inevitably translate into improved control strategies, resulting in increased food security and in maintaining the profitability of the poultry industry within a healthy environment for both humans and animals.

Joseph Domenech
Chief Veterinary Officer
Food and Agriculture Organization
of the United Nations (FAO)

Bernard Vallat
Director General
World Organisation for
Animal Health (OIE)

Introductory Remarks

This publication is a testimony to the efforts made by the staff of the International OIE/FAO Reference Laboratory (IRL) for Avian Influenza and Newcastle Disease, at the Istituto Zooprofilattico Sperimentale delle Venezie (IZSVe) in Padua, Italy, over the past few years in response to the global avian influenza crisis. Virologists, diagnosticians, molecular biologists and epidemiologists working at the IRL have assembled information collected in globally managing and diagnosing outbreaks not only of avian influenza but also of Newcastle disease, with the aim of improving animal and public health. We would like to express our sincere thanks to all members of the IZSVe staff involved in this project, to Dr. D.J. Alexander for his guidance and contributions and to Dr. B.Vallat and Dr. A. Petrini of the World Organisation for Animal Health (OIE) for their time and support. This publication would have not been possible without the support of the Food and Agriculture Organization of the United Nations (FAO), the European Commission, the Italian Ministry of Health and the Health and Veterinary Services of the Veneto Region, which have promoted and financially sustained the IRL through dedicated projects focusing on international collaboration. We are also very grateful to all of our international collaborators for supplying the figures and tables that make this publication unique.

Giuseppe Dalla Pozza
President
Istituto Zooprofilattico
Sperimentale delle Venezie

Igino Andrighetto
Director General
Istituto Zooprofilattico
Sperimentale delle Venezie

Stefano Marangon
Director of Science
Istituto Zooprofilattico
Sperimentale delle Venezie

cont.

I am honoured to be able to introduce this publication, as it is a tangible reflection of the levels of excellence reached by Italian scientists working in veterinary public health. The Italian, OIE and FAO Reference Laboratory for Newcastle Disease and Avian Influenza, based at the Istituto Zooprofilattico Sperimentale delle Venezie, Padua, Italy, is internationally recognised as one of the leading research and diagnostic laboratories in this field. It has led the way in the achievement of several breakthroughs in its areas of expertise, including vaccination strategies for the control and eradication of notifiable avian influenza, and in the creation of an international campaign, involving medical and veterinary research institutes, for sharing genetic data obtained from avian influenza isolates. In addition, the institute has supported and sustained diagnostic and research efforts throughout the world, particularly on the African continent, in Central Asia and in the Middle East, thereby generating data of relevance for the entire international scientific community in its efforts to manage the avian influenza threat. The global network established in response to the H5N1 crisis, and especially the cooperation of Mediterranean, African and Arab countries, has paved the way for productive collaborations in all aspects of veterinary public health and for continued progress in the overall objective of improving public health worldwide.

Romano Marabelli
Italian Chief Veterinary Officer

This publication follows a first volume on this topic, entitled "An Atlas and Text on Avian Influenza", published with the support of the Veneto Region in 2001. The first edition served to disseminate information collected by the staff of the Istituto Zooprofilattico Sperimentale delle Venezie during the Italian 1999–2000 H7N1 avian influenza epidemic—the forerunner to a series of devastating epidemics in Europe, the Americas, Asia and Africa. Ilaria Capua and her scientific team, in collaboration with Dennis Alexander, have collected data generated on a global level between 1999 and 2008 on avian influenza and Newcastle disease infections. Information on epidemiology, clinical signs, pathology, laboratory techniques and a vast collection of figures and tables have been assembled skilfully in this publication, aimed at supporting the efforts of diagnosticians, scientists and veterinary officers in their management of these infections.

We would like to join the authors in acknowledging the role of Dr. Giovanni Vincenzi, to whom this book is dedicated, for his pivotal role in managing the animal health crises that affected Northeastern Italy during his time in office.

We are grateful to the editors and authors of this publication for their efforts in this endeavour, which will certainly become an essential guide to combat avian influenza and Newcastle disease at a global level.

Giancarlo Galan
Governor, Veneto Region

Elena Donazzan
Chief Veterinary Health Authority
Veneto Region

Preface

In recent times, the worldwide spread of avian influenza (AI) viruses, particularly specific highly pathogenic AI viruses of H5N1 subtype, have put the livelihood of small rural poultry establishments, which historically had been threatened primarily by Newcastle disease (ND) viruses, at even greater risk. The occurrence of these two infections on a global scale is also threatening intensive poultry-farming systems and free-range establishments.

These diseases have several traits in common, including high flock mortality and certain clinical and pathological findings, and therefore may easily be misdiagnosed or confused with each other or with other viral or bacterial diseases.

In order to reduce the impact and spread of AI and ND, it is imperative that the disease is diagnosed properly and that appropriate measures are implemented to contain infection and, ultimately, eradicate the virus from an infected area. Both diseases have been shown to spread easily across boundaries and throughout entire continents. Thus, information on their epidemiology is essential to improve existing guidelines on their control.

This manual was conceived as a result of our efforts in directing international reference laboratories, thus gaining experience and information from outbreaks in many countries of the world. Our intention was to provide veterinarians and technicians with a field and laboratory manual containing all information relevant to the diagnosis and management of an AI or ND outbreak. In addition we have included information to support veterinary authorities in management issues.

The Cd-Rom that is included in this publication has been developed to supply trainers and university teachers with slides and videoclips of field and experimental infections; this will allow students and trainees to visualize the clinical and pathological traits of AI and ND. In addition PDF files of protocols and epidemiological inquiry forms may be downloaded for use.

We are confident that this publication will be useful to the different professions involved in the farming of birds. Ultimately we are convinced that through improved communication and diagnosis there will be a greater availability of information essential for an improved understanding of the ecology, epidemiology and animal and human health implications of these diseases.

Ilaria Capua *Dennis J. Alexander*

Acknowledgements

The editors and authors gratefully acknowledge the contribution of Giovanni Ortali, Veniero Furlattini, Anna Toffan, Roberta De Nardi, Ezio Bianchi, Francesco Prandini, Filippo Cilloni and of all those who have supplied the images included in this manual.

We are very thankful to Roberta Bassan, Anna Bidese, Marilena Campisi, Michaela Mandelli and Marta Vettore for their organisational and secretarial support.

Our special thanks are for Amelio Meini for his sketches and drawings and for the painting on the cover.

The editors and authors wish also to express their grateful thanks to the following colleagues for the permission to reproduce in the book scientific images and line-drawings of their property:

Nadim Mukhles Amarin Boehringer Ingelheim, Middle East Regional Office, Dubai, UAE

Daniel Baroux Laboratoire Départemental d'Analyses de l'Ain, Chemin de la Miche 01000 Bourg en Bresse, France

Caroline Brojer Department of Wildlife, Fish and Enviroment, National Veterinary Institute (SVA) SE-751 89 Uppsala, Sweden

Corrie Brown Department of Veterinary Pathology, College of Veterinary Medicine, 501 D.W. Brooks Drive, Athens, GA 30602-7388, Greece

Antonio Camarda Facoltà di Medicina Veterinaria, Dipartimento di Sanità e Benessere Animale, Sezione Patologie Aviarie, Str. Prov. per Casamassima Km 3, Valenzano, 70010 Bari, Italy

Ahmed Abd ElKarim Private Consultant, Giza, Cairo, Egypt

Victor Irza Federal Centre for Animal Health (FGI-ARRIAH), Russian Federation

Desiree Jansson Department of Poultry, SVA, 75189 Uppsala, Sweden

Walid Hamdy Kilany Animal Health Research Institute, National Lab. for Veterinary Quality Control on Poultry Production, Nadi El Saied st, Dokki, Giza, Egypt

Amelio Meini Intervet Italia, Via Tobagi 7, 20068 Peschiera Borromeo, Italy

Zenon Minta National Veterinary Research Institute, Al.Partyzantow, 57, Pulawy, PL-24-100, Poland

Vladimir Savic Director of the Croatian Veterinary Institute, Head of the Poultry Centre, Zagreb, Croatia

Thierry Van den Berg Veterinary and Agrochemical Research Center, Groeselenberg 99B-1180 Brussels, Belgium

A.H. Zahdeh Private Veterinarian, Jordan

Contents

1 **Ecology, Epidemiology and Human Health Implications of Avian Influenza Virus Infections** ... 1
Ilaria Capua and Dennis J. Alexander

2 **Ecology and Epidemiology of Newcastle Disease** ... 19
Dennis J. Alexander

3 **Notification of Avian Influenza and Newcastle Disease to the World Organisation for Animal Health (OIE)** ... 27
Antonio Petrini and Bernard Vallat

4 **Emergency Response on Suspicion of an Avian Influenza or Newcastle Disease Outbreak** ... 31
Manuela Dalla Pozza and Stefano Marangon

5 **Necropsy Techniques and Collection of Samples** ... 35
Calogero Terregino

6 **Clinical Traits and Pathology of Avian Influenza Infections, Guidelines for Farm Visit and Differential Diagnosis** ... 45
Ilaria Capua and Calogero Terregino

7 **Conventional Diagnosis of Avian Influenza** ... 73
Calogero Terregino and Ilaria Capua

8 **Molecular Diagnosis of Avian Influenza** ... 87
Giovanni Cattoli and Isabella Monne

9 **Clinical Traits and Pathology of Newcastle Disease Infection and Guidelines for Farm Visit and Differential Diagnosis** ... 113
Calogero Terregino and Ilaria Capua

10 **Conventional Diagnosis of Newcastle Disease Virus Infection** ... 123
Calogero Terregino and Ilaria Capua

11 **Molecular Diagnosis of Newcastle Disease Virus** ... 127
Giovanni Cattoli and Isabella Monne

12 **General Rules for Decontamination Following an Outbreak of Avian Influenza or Newcastle Disease** ... 133
Maria Serena Beato and Paola De Benedictis

Websites ... 151

Annex 1 **Check List for Visit to Suspect Premise** ... 153
Manuela Dalla Pozza

Annex 2 **Epidemiological Investigation Form for Avian Influenza and Newcastle Disease Outbreaks** ... 155
Manuela Dalla Pozza

Annex 3 **Biosafety Procedures** ... 169
William G. Dundon

Annex 4 **Laboratory Solutions** ... 175
William G. Dundon

Annex 5 **Guidelines for Shipping Avian Influenza and Newcastle Disease Virus Samples to OIE Reference Laboratories** ... 177
William G. Dundon

Subject Index ... 183

Contributors

Dennis John Alexander, OBE, BTech, PhD, CBiol FIBiol, FRCPath, DSc, Former Director of the EU OIE/FAO Reference Laboratory for Newcastle Disease and Avian Influenza, VLA Weybridge, KT15 3NB UK, d.j.alexander@vla.defra.gsi.gov.uk

Maria Serena Beato, DVM, OIE/FAO and National Reference Laboratory for Newcastle Disease and Avian Influenza, Virology Department; Istituto Zooprofilattico Sperimentale delle Venezie, Viale dell'Università 10, 35020 Legnaro, Padova, Italy, msbeato@izsvenezie.it

Ilaria Capua, DVM, PhD, Head of Virology Department, Director of OIE/FAO and National Reference Laboratory for Newcastle, Disease and Avian Influenza, Istituto Zooprofilattico Sperimentale delle Venezie, Viale dell'Università 10, 35020 Legnaro, Padova, Italy, icapua@izsvenezie.it

Giovanni Cattoli, DVM, PhD, Head of Research and Development Laboratory, Virology Department, OIE/FAO and National, Reference Laboratory for Newcastle Disease and Avian Influenza, Istituto Zooprofilattico, Sperimentale delle Venezie, Viale dell'Università 10, 35020 Legnaro, Padova, Italy, gcattoli@izsvenezie.it

Manuela Dalla Pozza, DVM, Epidemiplogy Department, Istituto Zooprofilattico, Sperimentale delle Venezie, Viale dell'Università 10, 35020 Legnaro, Padova, Italy, crev.mdallapozza@izsvenezie.it

Paola De Benedictis, DVM, OIE/FAO and National Reference Laboratory for Newcastle Disease and Avian Influenza, Virology Department, Istituto Zooprofilattico Sperimentale delle Venezie, Viale dell'Università 10, 35020 Legnaro, Padova Italy, pdebenedictis@izsvenezie.it

William G. Dundon, BA(mod), PhD, OIE/FAO and National Reference Laboratory for Newcastle Disease and Avian Influenza, Virology Department, Istituto Zooprofilattico Sperimentale delle Venezie, Viale dell'Università 10, 35020 Legnaro, Padova, Italy, wdundon@izsvenezie.it

Stefano Marangon, DVM, Director of Science, Istituto Zooprofilattico, Sperimentale delle Venezie, Viale dell'Università 10, 35020 Legnaro, Padova, Italy, smarangon@izsvenezie.it

Isabella Monne, DVM, OIE/FAO and National Reference Laboratory for Newcastle Disease and Avian Influenza, Research and Development Laboratory, Istituto Zooprofilattico Sperimentale delle Venezie, Viale dell'Università 10, 35020 Legnaro, Padova, Italy, imonne@izsvenezie.it

Antonio Petrini, DVM, Deputy Head Information Department, OIE, 12, Rue de Prony, 75017 Paris, France, a.petrini@oie.int

Calogero (Lillo) Terregino, DVM, PhD, Head of DiagnosticVirology, OIE/FAO and National Reference Laboratory for Newcastle, Disease and Avian Influenza, Istituto Zooprofilattico Sperimentale delle Venezie, Viale dell'Università 10, 35020 Legnaro, Padova, Italy, cterregino@izsvenezie.it

Bernard Vallat, DVM, Director General, OIE, 12, Rue de Prony, 75017 Paris, France, b.vallat@oie.int

Ecology, Epidemiology and Human Health Implications of Avian Influenza Virus Infections

1

Ilaria Capua and Dennis J. Alexander

1.1 Introduction

Avian influenza (AI) represents one of the greatest concerns for public health that has emerged from the animal reservoir in recent times. AI, in its highly pathogenic form (HPAI), has been known to the veterinary community since the end of the 19th century, when an Italian scientist, Edoardo Perroncito, reported what is believed to be the first documented evidence of "fowl plague" as a distinct disease. However, for over 100 years, HPAI proved to be a poultry disease of rare occurrence that, in most cases, affected an irrelevant number of birds. Generally speaking, it was either self-limiting or controlled efficiently through the application of measures aimed evadicating the infection from the affected area. At approximately the turn of the millennium, however, a sharp increase in the number of outbreaks of AI in poultry occurred. It has been calculated that the impact of AI on the poultry industry has increased 100-fold, with 23 million birds affected in the 40-year period between 1959 and 1998 and over 200 million from 1999 to 2004 (Capua and Alexander 2004). In addition, since 1997, the implications for human health of AI infections of poultry have been identified, especially as a result of the spread of Asian lineage HPAI H5N1 virus. This has dramatically attracted the attention of the scientific community and AI infections have assumed a completely different profile in both the veterinary and medical scientific communities.

In recent times, some outbreaks have remained of minor relevance while others, such as the Italian 1999–2000 H7N1, the Dutch 2003 H7N7 and the Canadian 2004 H7N3 outbreaks, have caused substantial damage to the poultry industry (Table 1.1). The Dutch and the Canadian outbreaks also resulted in human infections and thus caused more general human health concerns. Although these three outbreaks were believed to be exceptional with regards to magnitude, costs and human health involvement, they were merely a prelude to the spread of Asian H5N1 virus and the concerns generated about the emergence of a new pandemic virus for humans via the avian-human link.

1.2 Aetiology, Including the Emergence of HPAI

Influenza viruses have segmented, negative-sense, single-stranded RNA genomes and are placed in the Family *Orthomyxoviridae*. At present, the *Orthomyxoviridae* Family consists of five genera: *Influenzavirus A*, *Influenzavirus B*, *Influenzavirus C* (these genera were originally regarded as 'influenza types A, B and C' and this terminology is still in use), *Thogotovirus* and *Isavirus*. Only viruses of the *Influenzavirus A* genus are known to infect birds.

Type A influenza viruses are further divided into subtypes based on the antigenic relationships of the surface glycoproteins, haemagglutinin (HA) and neuraminidase (NA). To date, 16 HA subtypes (H1–H16) and nine NA subtypes (N1–N9) have been recognised. Each virus has one HA and one NA antigen, apparently in any combination. All influenza A subtypes in the majority of possible combinations have been isolated from avian species. Thus far, only viruses of H5, H7 and H10 subtypes have been shown to cause HPAI in susceptible species, but not all H5, H7 and H10 viruses are virulent.

For all influenza A viruses, the haemagglutinin glycoprotein is produced as a precursor, HA0, which requires post-translational cleavage by host proteases before it is functional and virus particles become infectious (Vey et al. 1992). The HA0 precursor proteins of avian influenza viruses of low virulence

I. Capua, D.J. Alexander (eds.) *Avian Influenza and Newcastle Disease*,

Table 1.1 Outbreaks of highly pathogenic avian influenza (HPAI) virus since 1959: causative agent, cleavage-site sequence, intravenous pathogenicity index (IVPI), number of farms, approximate number of birds, and other relevant features

	HPAI virus	Subtype	Amino acids at the HA0 cleavage site (/ = point of cleavage)	IVPI	Number of farms infected	Approximate numbers of poultry involved	Comments
1	A/chicken/Scotland/59	(H5N1)	PQRKKR/GLF	2.87	1 (small chicken farm)	NK	First H5 subtype influenza virus
2	A/turkey/England/63	(H7N3)	PETPKRRRR/GLF	2.78	2 (turkeys)	~29,000	
3	A/turkey/Ontario/7732/66	(H5N9)	PQRRKKR/GLF			8,000	
4	A/chicken/Victoria/76	(H7N7)	PEIPKKREKR/GLF	1.76		58,000	
5	A/chicken/Germany/79	(H7N7)	PEIPKKKKR/GLF, PEIPKRKKR/GLF, PEIPKKRKKR/GLF, PEIPKKKKKKR/GLF	NK	2 (1 chicken, 1 goose)	600,000 chickens 80 geese	Little information has been published on these outbreaks
6	A/turkey/England/199/79	(H7N7)	PEIPKKRKR/GLF, PEIPKRRRR/GLF, PEIPKKREKR/GLF	2.80	3 (turkeys)	9,262	Several isolates of LPAI H7 in the same vicinity, none with N7
7	A/chicken/Pennsylvania/1370/83	(H5N2)	PQKKKR/GLF	2.37		17,000,000	Low virulence H5N2 with same cleavage site circulated for 7 months before mutating to virulence
8	A/turkey/Ireland/1378/83	(H5N8)	PQRKRKKR/GLF	2.60 2.85	4 (2 turkey, 1 mixed, 1 duck)	306,020 (8,000 turkeys, 28,020 chickens, 270,000 ducks)	Surveillance showed commercial ducks to be infected, but no clinical signs were seen
9	A/chicken/Victoria/85	(H7N7)	PEIPKKREKR/GLF	2.80	1 (chicken)	240,000	HPAI confirmed on 1 farm but contacts slaughtered as a precaution
10	A/turkey/England/50-92/91	(H5N1)	PQRKRKTR/GLF	3.00	1 (turkey)	8,000	Low virulence virus (IVPI 0) with same HA0 cleavage site also isolated
11	A/chicken/Victoria/1/92	(H7N3)	PEIPKKKKR/GLF	2.71	1 (chicken)	18,000	H7 antibodies in healthy birds on the affected farm and neighbouring farms including a duck farm, all of which were slaughtered suggests LPAI virus infections prior to outbreak
12	A/chicken/Queensland/667-6/94	(H7N3)	PEIPRKRKR/GLF	2.87	1 (chicken)	22,000	

(continued)

Table 1.1 *(continued)*

13 A/chicken/Mexico/8623-607/94	(H5N2)	PQRKRKTR/GLF	NK	NK	NK	Situation confused due to presence of H5N2 LPAI virus, which continues to circulate, and use of vaccine
14 A/chicken/Pakistan/447/94	(H7N3)	PETPKRRKR/GLF PETPKRKRKR/GLF, PETPKRRNR/GLF	2.86	Many	3,200,000	H7N3 virus IVPI 2.8 re-emerged in 2003, its relationship to 1995 virus unclear. Situation complicated by widespread infections of LPAI H7N3 and H9N2 viruses
15 A/goose/Guandong/1/96 A/chicken/Hong Kong/97 A/chicken/Eurasia&Africa/2003	(H5N1)	PQRRRRKKR/GLF PQRRRRKKR/GLF PQRERRRKKR/GLF	2.1	See footnote	See footnote	Low virulence virus (IVPI 0) also isolated with same cleavage site motif A/goose/Guandong/2/96
16 A/chicken/NSW/97	(H7N4)	PEIPRKRKR/GLF	2.52, 2.90 1.30 (emus)	3 (2 chickens 1 emus)	160,000 chickens 261 emus	
17 A/chicken/Italy/330/97	(H5N2)	PQRRRKKR/GLF	3.00	8 (chickens, turkeys, geese, ducks, guinea fowl, pigeons, quail)	8,000	
18 A/turkey/Italy/99	(H7N1)	PEIPKGSRVRR/GLF	3.00	413	13,732,912	LPAI H7N1 circulated for 9 months before mutating to virulence
19 A/chicken/Chile/2002	(H7N3)	PEKPKTCSPLSRCRETR/GLF		2.43-3.00	~700,000	
20 A/chicken/Netherlands/2003	(H7N7)	PEIPKRRRR/GLF	2.94	255	>25,000,000a,b	
21 A/chicken/Texas/2004	(H5N2)	PQRKKR/GLF	0.00	1	6,600	HPAI because it has the same cleavage site as chicken/Scotland/59
22 A/chicken/Canada-BC/2004	(H7N3)	PENPKQAYRKRMTR/GLF	2.87		16,000,000	
23 A/ostrich/S. Africa/2004	(H5N2)	PQREKRRKKR/GLF	2.73		>30,000	Initial isolate an had IVPI of 1.19
24 A/chicken/N. Korea/2005	(H7N7)	PEIPKGRHRRPKR/GLF	3.00	3 (chicken)	219,000	

NK, not known.

The "Asian lineage HPAI H5N1 virus," for which A/goose/Guangdong/1/96 was the progenitor virus, has spread from southern China since 2003 and outbreaks have been reported throughout Asia and into Europe and Africa. The virus appears to have become endemic in poultry in some Asian, African and Middle Eastern countries.

[a]spread to Germany (one farm, 419,000 chickens) and Belgium (8 farms, 2,300,000 chickens).

[b]255 flocks were confirmed as infected, 233 commercial and 22 backyard or hobby flocks, in addition 1126 commercial and 16,521 hobby flocks were culled pre-emptively

(LPAI viruses) for poultry have a single arginine at the cleavage site and another basic amino acid at position -3 or -4 from the cleavage site. These viruses are limited to cleavage by extracellular host proteases such as trypsin-like enzymes and thus restricted to replication at sites in the host where such enzymes are found, i.e. the respiratory and intestinal tracts. H5 and H7 HPAI viruses possess multiple basic amino acids (arginine and lysine) at their HA0 cleavage sites, either as a result of apparent insertion or apparent substitution (Wood et al. 1993; Senne et al. 1996; Vey et al. 1992), and are cleavable by one or more intracellular ubiquitous proteases, probably one or more proprotein-processing subtilisin-related endoproteases of which furin is the leading candidate (Stieneke-Gröber et al. 1992). H5 and H7 HPAI viruses are able to replicate throughout the bird, damaging vital organs and tissues, which results in disease and death.

Some H10 viruses fall within the definition of HPAI (Alexander 2008) but these viruses do not have a multi-basic cleavage site and do not cause systemic infection. They are known to have tropism for the kidney, and the impaired function of this organ is what determines death of the bird when the virus is administered intravenously (Swayne and Alexander 1994).

The converse is also true and, as reviewed by Londt et al. (2007), at least four viruses with unusual virulence characteristics have been isolated: A/chicken/Pennsylvania/1/83 (H5N2), A/goose/Guangdong/2/96 (H5N1), A/turkey/England/87-92BFC/91 (H5N1) and A/chicken/Texas/298313/04 (H5N2). These have multiple basic amino acid motifs at the HA0 cleavage site, similar to known virulent AI viruses, but do not show virulence for chickens.

To date, apart from the H10 isolates, only viruses of the H5 and H7 subtypes have been shown to cause HPAI. It appears that most of these HPAI viruses arose by mutation after LPAI H5 and H7 viruses were introduced into poultry from the wild bird reservoir. Several mechanisms seem to have been responsible for this mutation. Most HPAI viruses appear to have arisen as a result of spontaneous duplication of purine triplets, which resulted in the insertion of basic amino acids at the HA0 cleavage site; this most likely occurred due to a transcription error by the polymerase complex (Perdue et al. 1998). However, as pointed out by Perdue et al., this is clearly not the only mechanism by which HPAI viruses can arise, as some have resulted from nucleotide substitution rather than insertion while others have insertions without repeating nucleotides. The Chile 2002 (Suarez et al. 2004) and Canada 2004 (Pasick et al. 2005) H7N3 HPAI viruses show distinct and unusual cleavage-site amino acid sequences. These viruses arose as a result of recombination with other genes (nucleoprotein gene and matrix gene, respectively), resulting in an insertion at the cleavage site of 11 amino acids for the Chile virus and 7 amino acids for the Canadian virus.

The factors that bring about mutation from LPAI to HPAI are not known. In some instances, mutation seems to have taken place rapidly (at the index case site) after introduction from wild birds; in others, the LPAI virus progenitor circulated in poultry for months before mutating. Therefore, it is impossible to predict if and when this mutation will occur. However, it can be reasonably assumed that the wider the circulation of LPAI in poultry, the higher the chance that there will be a mutation to HPAI. In some cases, LPAI viruses of the H5 or H7 subtype circulated for very long periods of time without mutating to the highly pathogenic form (Davison et al. 2003; Senne et al. 2006; Senne 2007).

There is some field and laboratory evidence consistent with the emergence of HPAI after introduction into poultry. For example, results of phylogenetic studies of H7 subtype viruses indicated that HPAI viruses do not constitute a separate phylogenetic lineage or lineages, but appear to have arisen from nonpathogenic strains (Banks et al. 2000; Röhm et al. 1995). This is supported by the in vitro selection of mutants virulent for chickens from an avirulent H7 virus (Li et al. 1990). An opportunity to examine the genetic changes leading up to the appearance of HPAI virus came with the LPAI and HPAI H7N1 outbreaks in Italy in 1999–2000, in which 199 LPAI outbreaks preceded a mutation to virulence, which then resulted in 413 HPAI outbreaks and the loss of some 16,000,000 birds (Capua and Marangon 2007). Phylogenetic analyses of the LPAI and HPAI viruses suggested that the outbreaks occurred as a result of a single introduction of LPAI virus, with mutation to virulence occurring in a single lineage, probably driven by evolutionary processes associated with the adaptation of the virus to a new host (poultry) (Capua et al. 2000).

HPAI viruses are not necessarily virulent for all species of birds and the clinical severity seen in any host varies with both bird species and virus strain (Alexander 2000). In particular, ducks, prior to the emergence of the Asian H5N1, rarely showed clinical signs as a result of HPAI infections. Ostriches also have an atypical clinical response to AI virus infections, as LPAI and HPAI viruses seem to cause a similar clinical condition, with adult birds showing some resistance, even to HPAI, while young birds show similar clinical signs and mortality with both pathotypes (Capua and Mutinelli 2001).

1.3 Host Range

Avian influenza viruses have been shown to infect birds and mammals. Generally speaking, the former are infected more readily and efficiently than the latter, and the interspecies and intraspecies transmission within the Class *Aves* occurs to a greater extent than in the Class *Mammalia*. One of the main factors that influence susceptibility to infection is the receptor conformation on the host cells. AI viruses bind preferably to sialic acid (SA)-α2,3-Gal-terminated saccharides, which are prominent on avian cells. Human influenza viruses, in contrast, bind preferentially to SA-α2,6-Gal-terminated saccharides, well-represented on human epithelial cells. This different binding preference is believed to be one of the major factors that impede crossing of the species barrier. However, the fact that AI viruses do occasionally infect people and other mammals indicates that this barrier is not insurmountable.

1.3.1 Birds

Influenza viruses have been shown to infect a great variety of birds (Alexander 2000, 2001; EFSA 2005; Hinshaw et al. 1981a; Lvov 1978), including free-living birds, captive caged birds and domestic ducks, chickens, turkeys and other domestic poultry.

It was not until the mid-1970s that systematic investigations of influenza in feral birds were undertaken. These investigations revealed enormous pools of influenza viruses to be present in the wild bird population (EFSA 2005; Olsen et al. 2006; Stallknecht 1988; Stallnecht and Shane 1998), especially in waterfowl, *Family Anatidae*, *Order Anseriformes*. In the surveys listed by Stallknecht and Shane (1998), a total of 21,318 samples from all species resulted in the isolation of 2317 (10.9%) viruses. However, 14,303 of these samples were from birds of the Order Anseriformes, which yielded 2173 (15.2%) of the isolates. The next highest isolation rates were 2.9 and 2.2%, from the *Passeriformes* and *Charadriiformes*, respectively; but these compare with an overall isolation rate of 2.1% from all birds other than ducks and geese. However, studies by Sharp et al. (1993) suggested that waterfowl do not act as a reservoir for all AI viruses. It seems likely that part of the influenza gene pool is maintained in shorebirds and gulls, from which the predominant number of isolated influenza viruses are of a subtype different from those isolated from ducks (Kawaoka et al. 1988).

Prior to the ongoing H5N1 epizootic, HPAI had only once affected wild birds significantly. This outbreak occurred in South Africa in 1961 and caused the death of approximately 1300 common terns (Becker 1966). It appeared, therefore, that HPAI was a disease of domesticated birds and that wild birds usually only harboured the low pathogenic form of these viruses. The unprecedented situation occurring in Asia has resulted in the spill-over of infection to naïve populations of wild birds. In particular, it was suggested that the presence of the virus in migratory birds at Lake Qinghai, in Western China, resulting in the death of many bar-headed geese, could have been the means by which the H5N1 virus spread west and south (Chen et al. 2005; Liu et al. 2005). However, nearly all the wild birds from which Asian HPAI H5N1 virus has been isolated were either dead or dying. In addition, the incubation period of this disease in migratory birds is unknown and probably shows considerable variability among taxonomic families and species. In very simple terms, at the moment, the scientific community only has an indication of the species that may be infected and succumb to the virus. Knowledge and information on all species that are susceptible to infection, including the incubation period for those birds that do develop a clinical condition, their ability to fly significant distances if infected and data on the route, duration and titre of viral shedding, are unavailable. Some reports, such as the isolation of HPAI H5N1 from apparently healthy individuals from three species of migratory birds on Lake Qinghai one year after the first detection of the virus there (Lei et al. 2007), are of concern. At this stage, only hypotheses can be formulated on the eco-epidemiological consequences of this spill-over of HPAI virus into wild birds.

However, ongoing surveillance efforts in Eurasian and African wild bird populations suggest that the HPAI H5N1 virus is currently not truly endemic in wild birds, and that the occurrence of H5N1 in wildfowl in Europe and in the Mediterranean in 2006 and 2007 was probably linked to an exceptional situation—and can most probably be considered as having been a self-limiting event. The continuation of surveillance efforts in collaboration with ornithologists will generate additional data to draw final conclusions on this aspect.

The Asian H5N1 viruses and their descendants have infected a variety of wild birds in Europe, the majority of which were waterfowl. Among these, mute swans (*Cygnus olor*) appeared to have been the most affected species (Alexander 2007).

1.3.2 Avian Influenza Infections of Mammals

1.3.2.1 Pigs

Ostensibly, pigs play a crucial role in influenza ecology and epidemiology, primarily because of their dual susceptibility to human and avian viruses. They possess both SA-α2,3-Gal-terminated saccharides and SA-α2,6-Gal-terminated saccharides and are therefore considered a potential "mixing vessel" for influenza viruses, from which reassortants may emerge. Kida et al. (1994) demonstrated experimentally that pigs were susceptible to infection by at least one virus representative of each of the subtypes H1–H13.

The introduction of classical swine H1N1 influenza viruses to turkeys from infected pigs has been reported to occur regularly in the USA and, in some cases, influenza-like illness in pigs has been followed immediately by disease signs in turkeys (Halvorson et al. 1992; Mohan et al. 1981; Pomeroy 1982). Genetic studies of H1N1 viruses from turkeys in the USA have revealed a high degree of genetic exchange and reassortment of influenza A viruses from turkeys and pigs in the former species (Wright et al. 1992). In Europe, avian H1N1 viruses were transmitted to pigs, became established and were subsequently reintroduced to turkeys from pigs (Ludwig et al. 1994; Wood et al. 1997). An independent introduction of H1N1 virus from birds to pigs occurred in Europe in 1979 (Pensaert et al. 1981). A similar introduction occurred in Asia in the early 1990s; these latter viruses are genetically distinct from the viruses in Europe (Guan et al. 1996). H9N2 viruses were introduced into pigs in South-East Asia (Peiris et al. 2001). Serological evidence has been obtained of infections of pigs with viruses of H4, H5 and H9 subtypes (Karasin et al. 2004). During the HPAI H7N7 epidemic in The Netherlands, in 2003, 13 pig herds on farms with infected poultry were shown to have antibodies to the H7 subtype, although no virus was detected (Loeffen et al. 2003, 2004). In Canada, however, avian viruses of H3N3 and H4N6 subtypes have been isolated from pigs (Karasin et al. 2000, 2004). Clearly, the introduction of AI viruses to pigs is not an uncommon occurrence. Nonetheless, the only subtypes to have become truly established in pig populations and readily transmissible from pig-to-pig and herd-to-herd are H1N1, H3N2 and the reassortant H1N2, although genotype analysis of isolates of these subtypes suggests that they can be the result of reassortment of viruses from different progenitor host species (pig, human and avian).

There have been sporadic unpublished reports of natural infection of H5N1 in pigs (Van Reeth 2007), and three experimental infections have been carried out with a total of eight H5N1 viruses. Six of these replicated in pigs, although clinical signs were mild or unapparent and shedding levels were not high. In none of the experiments was pig-to-pig transmission observed (Choi et al. 2005; Isoda et al. 2006; Shortridge et al. 1998).

1.3.2.2 Horses

Although there have been isolated reports of evidence of infection of horses with viruses of subtypes H1N1, H2N2 and H3N2 (Tůmová 1980), influenza infections of horses have been restricted essentially to H7N7 and H3N8 subtypes of influenza A; these viruses form distinct lineages in phylogenetic studies. However, examination of H3N8 viruses isolated from severe epidemics in horses occurring in the Jilin and Heilongjiang Provinces in the northeast of the People's Republic of China in 1989 and 1990 showed them to be antigenically and genetically distinguishable from other equine H3N8 viruses. Thus, Guo et al. (1992) concluded that this virus was of recent avian origin and had probably spread directly to horses without reassortment. This virus does not appear to have become established in the horse population. To date, there is no evidence that H5N1 has infected horses naturally.

1.3.2.3 Marine Mammals

During 1979 and 1980, approximately 500 deaths, corresponding to about 20% of the population, occurred in harbour seals (*Phoca vitulina*) around the Cape Cod Peninsula in the USA as a result of acute haemorrhagic pneumonia. Influenza A viruses of H7N7 subtype were isolated repeatedly from the lungs or brains of the dead seals (Lang et al. 1981). The virus infecting the seals was shown to be closely related both antigenically and genetically to AI viruses (Webster et al. 1981) and may have represented direct transmission to the seals without reassortment.

In 1983, further deaths (2–4%) occurred in harbour seals on the New England coast of the USA and influenza A virus of subtype H4N5 was isolated. Once again, all eight genes of this virus were demonstrably of avian origin (Webster et al. 1992). Following the surveillance of seals on the Cape Cod Peninsula, two influenza A viruses of the H4N6 subtype were isolated in 1991 and three of H3N3 subtype in 1992—all from seals found dead with apparent viral pneumonia (Callan et al. 1995). Antigenic and genetic characterisation revealed that these too were avian viruses that had entered the seal population.

Two viruses, of H13N2 and H13N9 subtypes, were isolated from a single beached pilot whale; genetic analysis indicated that the viruses had been introduced recently from birds (Chambers et al. 1989; Hinshaw et al. 1986).

1.3.2.4 Mustelids

In October 1984, outbreaks of respiratory disease affected approximately 100,000 mink on 33 farms situated in close proximity along a coastal region of southern Sweden, with 100% morbidity and 3% mortality (Klingeborn et al. 1985). Influenza A viruses of H10N4 subtype were isolated from the mink; genetic analysis indicated that the viruses were of avian origin and very closely related to a virus of the same subtype isolated from chickens and a feral duck in England in 1985 (Berg et al. 1990). Earlier experimental infections had suggested that mink were susceptible to infection with various subtypes of AI viruses (Okazaki et al. 1983).

An Asian lineage H5N1 HPAI virus was isolated from a wild stone marten (*Martes foina*) found sick in Germany on 2 March 2006, a time when there had been numerous reports of H5N1 virus in wild birds in the area where the marten was found (WHO http://www.who.int/csr/don/2006_03_09a/en/index.html).

1.3.2.5 Felids

Studies by Hinshaw et al. (1981b) had shown the ability of LPAI viruses to infect and replicate in cats, which did not exhibit subsequent clinical signs of infection. However, during the 2003–2004 HPAI H5N1 outbreak in Asia, there were occasional reports of fatal H5N1 virus infections in domestic cats and zoo felids after they had been fed virus-infected chickens (Keawcharoen et al. 2004; Songserm et al. 2006; Thanawongnuwech et al. 2005). Mortality in cats caused by a clade 2.2 Qinghai Lake lineage virus was also reported in Iraq (Yingst et al. 2006). In experimental studies, cats excreted virus and developed lung pathology after intratracheal inoculation with H5N1 or after feeding on H5N1-virus-infected chickens (Kuiken et al. 2004). In addition, the virus was transmitted from infected to sentinel cats. Thus, cats may become infected with AI after consumption of fresh infective poultry meat and they may spread the virus to other cats.

1.3.2.6 Dogs

Dogs were believed to be resistant to AI infection until a recent case of H5N1 in Thailand. A fatal infection was caused most probably by ingestion of an infected duck carcass. H5N1 was recovered from the dog's lungs, liver and kidney and from urine specimens (Songserm et al. 2006).

1.4 Transmission

The mechanisms by which influenza viruses pass from one bird to another and bring about infection are poorly understood. In the past, attempts were made to assess the transmissibility of LPAI and HPAI viruses in domestic poultry experimentally (Alexander et al. 1978, 1986; Narayan et al. 1969; Tsukamoto et al. 2007; Westbury et al. 1979, 1981). The results suggested that bird-to-bird transmission is extremely complex and depends on the strain of virus, the species of bird and environmental factors.

In both natural and experimental infections, virulent viruses have tended to show much poorer transmission from infected to susceptible chickens and turkeys than viruses of low pathogenicity. The ability of virus to spread easily must, to some extent, be related to the amount of virus released by the respiratory or intestinal route. The HPAI viruses cause extremely rapid death in these birds and it is possible that relatively little virus is excreted during the course of such infections.

The different epidemiology of the Asian H5N1 HPAI has led to several research groups re-examining the mechanism of AI virus transmission. In particular, the change in transmission from primarily the faecal/oral route to primarily the respiratory route in land birds, especially in minor poultry species such as quail and pheasants, has been considered significant in the epidemiology of that virus, particularly in its spread to mammals (Perez et al. 2003; Makarova et al. 2003; Humbrerd et al. 2006).

In any case, it seems that transmission from bird to bird occurs as a result of close proximity between infected and naïve hosts. Generally speaking, it is believed that direct contact with infected birds or with contaminated exudates or droppings are necessary for infection to be transmitted from one bird to another. This also indicates that airborne spread over large distances is an unlikely event. At present, there is no published evidence of such an occurrence.

1.5 Distribution and Spread

Until recently, it appeared that the epidemiology of AI consisted of the perpetuation of LPAI viruses of all H subtypes in wild birds, in which they caused little or no disease, with spread from time to time to poultry. Rarely, introductions of LPAI viruses of H5 or H7 subtype into poultry resulted in the mutation of these viruses to virulent viruses that caused HPAI.

The degree to which LPAI or HPAI viruses occur and spread in poultry may vary considerably and depends on the levels of biosecurity and the concentration of poultry in the vicinity of the initial outbreaks or the emergence of HPAI virus. However, events during the late 1990s and especially after 2003 have completely changed our understanding of AI epidemiology. The spread of LPAI virus of H9N2 subtype and HPAI virus of H5N1 subtype need separate consideration from the more conventional situation.

1.5.1 Conventional Situation

1.5.1.1 Primary Introduction to Poultry

All available evidence suggests that in the conventional situation the *primary* introduction of LPAI viruses into a poultry population is a result of wild bird activity, usually waterfowl, but gulls and shorebirds have also been implicated. This may not necessarily involve direct contact, as infected waterfowl may carry the viruses into an area and these may then be introduced to poultry by humans, other types of birds or other animals, none of which need to be infected but which may transfer the virus mechanically in infective faeces or exudates from the waterfowl. Surface water used for drinking water may also be contaminated with influenza viruses and thus serve as a source of infection. There is much evidence implicating waterfowl in the vast majority of primary LPAI outbreaks. In summary, this evidence is as follows:

- There is a much higher prevalence of infection of poultry on migratory waterfowl routes, although in view of the variation in virus excretors along the flyways (Pomeroy 1982) this may occur more frequently at some stages of the migratory route than others, e.g. in Minnesota, USA, compared to other states on the Mississippi flyway (Pomeroy 1982), and in Italy (Terregino et al 2007).
- There is a higher prevalence of infection of poultry kept in exposed conditions (e.g. turkeys on range, ducks on fattening fields). Conversely, where there have been regular LPAI infections and change to a policy of confinement has thus been pursued LPAI problems have largely disappeared (Lang 1982; Pomeroy 1987).
- Surveillance studies in areas with LPAI problems in poultry have shown the same variation in virus subtypes in sampled waterfowl and turkey outbreaks (Senne 2003). Similarly, backyard flocks reared in the open in Italy have been shown to harbour the same viruses isolated from waterfowl during the same period (Terregino et al 2007).
- Influenza outbreaks show a seasonal occurrence in high-risk areas, which coincided with migratory activity (Halvorson et al. 1983, 1987).

In most documented specific outbreaks, evidence has been obtained of probable waterfowl contact at the initial site.

On some occasions, primary introduction to poultry has resulted from a commercial sector where AI virus may be endemic. A good example of this is the H7N2 LPAI outbreaks in the USA (Halvorson 1987). LPAI virus of H7N2 subtype was probably introduced into live-bird markets in the eastern USA in 1994. Despite attempts to eradicate the virus, it has remained endemic since then. Senne et al. (2006) reported that, in the last 10 years, eight LPAI H7N2 outbreaks in commercial poultry, resulting in the slaughter of millions of birds and severe economic losses, have been linked to the live bird markets.

1.5.1.2 Secondary Spread

The greatest threat of spread of AI viruses is by mechanical transfer of infective organic material. In faeces, for example, the virus may be present at concentrations as high as 10^7 infectious particles/g and may survive for longer than 44 days (Utterback 1984). Birds or other animals that are not themselves susceptible to infection may become contaminated and spread the virus mechanically. Shared water or food may also become contaminated. However, for domestic poultry the main source of secondary spread appears to be humans. In several specific accounts, strong evidence has implicated the movements of caretakers, farm owners and staff, trucks and drivers moving birds or delivering food and artificial insemination crews in the spread of the virus both onto and through a farm (Glass et al. 1981; Homme et al. 1970; Wells 1963).

Spread by personnel and fomites was the method most strongly suspected in the widespread and devastating H5N2 epizootic in chickens in Pennsylvania during 1983–1984. Although there was some evidence that windborne spread played a role amongst very closely situated farms and that flying insects could become contaminated with infected faeces, it was concluded by most observers that secondary spread was principally due to the movement of personnel and equipment between farms (Johnson 1984; King 1984; Utterback 1984). King (1984) listed six types of fomite that may be moved from farm to farm and 11 types of personnel that may be in contact with two or more farms; Utterback (1984) produced even longer lists. In more recent outbreaks, such as those in Italy in 1999–2000, the density of the poultry population in the infected area and the frequent contact between farms by feed trucks, abattoir trucks and other vehicles have been associated with the spread of virus (Capua et al. 2001).

1.5.2 H9N2 Virus in Poultry

Historically, LPAI viruses have not been the subject of notification and control aimed at eradication. Thus, it was not clear why these viruses had not become more ubiquitous and endemic in poultry across large geographical areas, as had other viruses such as avian pneumoviruses or avian infectious bronchitis viruses. However, this is exactly what seems to have occurred with H9N2 LPAI viruses, and infections of poultry, mainly chickens, have occurred in many countries since the mid-1990s, reaching panzootic proportions. Outbreaks due to H9N2 AI occurred in domestic ducks, chickens and turkeys in Germany during 1995–1997, 1998 and 2004 (Werner 1998, 1999); in chickens in Italy in 1994 and 1996 (Fioretti et al. 1998), pheasants in Ireland in 1997 (Campbell 1998), ostriches in South Africa in 1995 (Banks et al. 2000), turkeys in the USA in 1995 and 1996 (Halvorson et al. 1998) and in chickens in Korea in 1996 (Mo et al. 1998). More recently, H9N2 infections have been reported in the Middle East and Asia, causing widespread outbreaks in commercial chickens in Iran, Saudi Arabia, Pakistan, China, Korea, UAE, Israel, Jordan, Kuwait, Lebanon, Libya and Iraq (Alexander 2002, 2007). In several of these countries, vaccine has been deployed to bring the disease under control. Nevertheless, H9N2 infections have become endemic in commercial poultry in a significant number of countries.

1.5.3 Spread of Asian HPAI H5N1 Virus

The emergence of HPAI H5N1 virus in Southeast Asia and its spread across Asia and into Europe and Africa is unprecedented in the virological era. The apparent progenitor virus for the subsequent outbreaks of HPAI of H5N1 subtype was obtained from an infection of commercial geese in Guandong province, People's Republic of China, in 1996 (Xu et al. 1999). In some reports, it has been considered that the virus continued to circulate in southern China primarily in domestic ducks and showed some genetic variation (Sims et al. 2005). This apparent low-level, but probably endemic, situation changed dramatically in December 2003 to February 2004, when suddenly eight countries in East and Southeast Asia reported outbreaks of HPAI due to H5N1 virus (Sims et al. 2005). Although there seemed to be some success in controlling the outbreaks in some countries, it re-emerged in a second wave beginning in July 2004. Malaysia reported an outbreak in poultry in August 2004 and became the ninth country in the region to be affected (OIE 2006). The virus affected all sectors of the poultry populations in most of these countries, but its presence in free-range commercial ducks, village poultry, live bird markets and fighting cocks seemed especially significant in the spread of the virus (Sims et al. 2005; Xu et al. 1999; Songserm et al. 2006).

If HPAI virus becomes widespread in poultry, particularly in domestic ducks that are reared on free range, spill-over into wild bird populations is inevitable. In the past, such infections have been restricted to wild birds found dead in the vicinity of infected poultry, but there has always been concern that infections of wild birds in which HPAI virus caused minimal or no clinical signs (i.e., ducks) could result in spread of the virus over large areas and long distances. Outbreaks affecting many wild bird species at two waterfowl parks in Hong Kong were recorded in 2002 (Ellis et al. 2004) and further, possibly more significant, outbreaks in wild migratory birds were reported in China and Mongolia in 2005. The presence of HPAI H5N1 virus in migratory birds at Lake Qinghai in Western China (Chen et al. 2005; Liu et al. 2005) was of particular concern for the spread of the virus.

There is no substantiated evidence that wild birds were responsible for the introduction of virus into Russia but HPAI H5N1 virus, genetically closely related to isolates obtained at Lake Qinghai, reached poultry there in the summer of 2005. These Russian viruses may have been the progenitors of viruses that spread further west between the end of 2005 and the beginning of 2006 (Salzberg et al. 2007). It is not clear whether westward spread was associated with movements of poultry or wild birds, probably both were involved; but during 2005 to the beginning of 2006, genetically closely related H5N1 viruses appeared in a number of countries in the region. The derivatives could be further divided into three sublineages, EMA1, EMA2 and EMA3 (Salzberg et al. 2007).

Reports of HPAI H5N1 virus infections continued in the first three months of 2006 and by early April 2006 31 countries from Asia, Europe and Africa had reported HPAI caused by H5N1 virus to the World Organisation for Animal Health (OIE) since the end of 2003 (OIE 2006).

Two isolated incursions of HPAI H5N1 virus into Europe occurred in 2004 and 2005 and are good examples of the influence of humans in the potential spread of AI viruses. In the first, crested-hawk eagles (*Spizaetus nipalensis*) smuggled from Thailand were confiscated at Brussels Airport, Belgium, and shown to be infected with H5N1 virus genetically similar to viruses isolated in Thailand (Van Borm et al. 2005). In the second, investigations of deaths in captive caged birds held in quarantine in England, ostensibly originating from Taiwan, showed them to be a result of HPAI H5N1 infection (DEFRA 2005). In this case, the virus was genetically closest to viruses isolated in China.

Isolates from dead swans were obtained in Croatia in October 2005 (OIE 2006). These infected swans were a forerunner of the apparent importance of these birds in the spread of HPAI H5N1. During January to April 2006, wild mute swans or other wild birds in Azerbaijan, Iran, Kazakhstan, Georgia and 20 European countries were shown to be infected. What appeared to have occurred was that mute swans or other birds over-wintering on the Black Sea became infected at a time when adverse weather conditions made the Black Sea inhospitable and the birds therefore dispersed to other areas. However, this would not explain the appearance of apparently the same H5N1 strain in swans and wild birds on the Baltic Coast at the same time.

Although reports of outbreaks in poultry in Asia and Africa continued during the winter of 2006–2007, in Europe these ceased, and surveillance studies failed to detect virus in wild birds or poultry. However, in January 2007, there were outbreaks in geese in Hungary and then on a turkey farm in England. The viruses isolated at the two sites were closely related and similar to viruses isolated from wild birds in Europe in 2006 (Irvine et al. 2007). Nevertheless, the viruses responsible for these outbreaks appeared to be distinguishable from those causing outbreaks in wild birds in Germany, Czech Republic (where outbreaks were also reported in poultry) and France during June and July 2007.

Following the first incursion of H5N1 in Africa, between early 2006 and late 2007, the virus spread to ten African countries (OIE 2007). In early 2007, an investigation was carried out to establish the genetic characteristics of contemporary H5N1 strains co-circulating in Nigeria. It appeared that, of the original sublineages introduced in the continent, only two (EMA 1 and EMA2) were still circulating, although an EMA1/EMA2 reassortant had become predominant (Monne et al. 2007).

1.6 HPAI Outbreaks

The outbreaks of HPAI in poultry since 1959 (when the first known HPAI outbreak due to virus of H5 subtype occurred) are listed in Table 1.1. If the Asian H5N1 HPAI virus outbreaks are considered to be a single epizootic, then there have been 24 outbreaks or epizootics in that time. Table 1.1 gives a summary of the basic information on HPAI outbreaks and points out how diverse such outbreaks may be. It can also be noted that, in the first 34 years of the 47-year period (1959–1992), there were 11 outbreaks (frequency 1 outbreak/3.09 years). In the next 13 years, there were another 13 (frequency 1 outbreak/1.0 year), while in the last 5 years (2002–2006) there have been six outbreaks (frequency 1 outbreak/0.83 years), including the unprecedented spread of the Asian H5N1 virus. Perhaps even more alarming than the increase in HPAI outbreaks is the number of birds affected. While in the first 12 outbreaks only one resulted in more than 500,000 birds dying or being slaughtered, deaths in eight of the second 12 greatly exceeded 500,000 birds (Table 1.1).

The reasons for the apparent increase in both numbers of outbreaks and their impact are likely to be extremely complex and a product of several factors: greater awareness and diagnostic capabilities; changes in poultry production such as establishing densely populated poultry production areas, integrated production systems and a move towards rearing birds on open range; more open reporting and investigation of disease; and, possibly, changes in wild bird movements as a result of climate change.

1.7 Human Health Implications

Due to the recent cases of human infection caused by AI viruses and to concern about the generation of a new pandemic virus originating from the H5N1 virus, AI infections are now considered a significant threat for public health.

Although it has been known for some time that the human pandemic viruses of 1957 and 1968 appeared to arise by reassortment between viruses present in the human population and AI viruses (Gething et al. 1980; Kawaoka et al. 1989; Scholtissek et al. 1978), because of the apparent "barriers" to human influenza viruses infecting birds and AI viruses infecting humans, it was suggested that pigs, which both human and avian viruses are known to infect readily, acted as "mixing vessels." The scientific basis for this was that pig epithelial cells contain both SA-α2,3-Gal-terminated saccharides and SA-α2,6-Gal-terminated saccharides, and this allows the replication of both avian and human influenza viruses. Reassortment between human and avian influenza viruses could therefore take place in pigs, with the emergence of viruses with the necessary gene(s) from the virus of human origin to allow replication and spread in the human population, but with a different haemagglutinin surface glycoprotein, so that the human population could be regarded as immunologically naïve.

However, recent events have changed significantly our understanding of infections of humans with AI viruses. A summary of reported cases is presented in Table 1.2. As indicated, until 1996 there had been only three reported infections and these had been the result of unknown contact, in 1959, and two laboratory accidents, in 1977 and 1981 (with the seal isolate). This was in keeping with the findings of Beare and Webster (1991), i.e. that, in experiments, human volunteers produced, at best, only transitory infections when challenged with AI viruses.

The first reported infection of a human known to have had contact with birds was the isolation of an avian virus of H7N7 from a woman in England who kept ducks and presented with conjunctivitis (Banks et al. 1998; Kurtz et al. 1996). This was the vanguard of the series of isolations from people having contact with poultry shown in Table 1.2. The impact of these subsequent human infections on public health issues was greatly enhanced by the high death rate in those confirmed to be infected. Deaths usually occurred as a result of severe respiratory disease and, although there were other symptoms, there was usually no evidence that virus replicated outside the respiratory tract (Yuen et al. 1998) and the infections were not comparable to the systemic infections seen in poultry.

The biggest threat resulting from of the demonstration of direct natural infections of humans with AI viruses is that pandemic viruses could emerge as a result, without an intermediate host. There are two mechanisms by which this could occur: by genetic reassortment or by progressive adaptation. The first case would occur if a person was simultaneously infected with an AI virus and a "human" influenza virus. In this case, through genetic reassortment, a virus fully capable of spread in the human population, but with H5, H7 or H9 haemagglutinin could emerge, resulting in a true influenza pandemic. However, it seems likely that during the widespread outbreaks of H9N2 virus since the mid-1990s and the H5N1 outbreaks in Asia since 1996 many more people than those listed in Table 1.2 could have been infected with

Table 1.2 Cases of human infection caused by avian influenza viruses (1959 to September 30, 2007)

Year	Country	Subtype	Number infected	Number deaths	Symptoms	Reference
1959	USA	H7N7	1	0	Hepatitis?	Campbell et al. (1970)
1977	Australia	H7N7	1	0	Conjunctivitis	Taylor and Turner (1977)
1981	USA	H7N7	1	0	Conjunctivitis	Webster et al. (1981)
1996	England	H7N7	1	0	Conjunctivitis	Kurtz et al. (1996)
1997	China	H5N1	18	6	Influenza-like illness	Chan et al. (2002)
1999	China	H9N2	2	0	Influenza-like illness	Peiris et al. (1999)
2002	USA	H7N2	1	0	Serological evidence	CDC website
	China	H5N1	2	1	Influenza-like illness	CDC website
		H9N2	1	0	Influenza-like illness	Butt et al. (2005)
2003	The Netherlands	H7N7	89	1	Conjunctivitis Influenza-like illness	CDC website
	USA	H7N2	1	0	Influenza-like illness	CDC website
	Vietnam	H5N1	3	3	Influenza-like illness	WHO website
	China	H5N1	1	1	Influenza-like illness	WHO website
	Italy	H7N3	7	0	Serological evidence	Puzelli et al. (2005)
2004	Canada	H7N3	2	0	Influenza-like illness	CDC website
	Thailand	H5N1	17	12	Influenza-like illness	WHO website
	Vietnam	H5N1	29	20	Influenza-like illness	WHO website
2005	Cambodia	H5N1	4	4	Influenza-like illness	WHO website
	China	H5N1	8	5	Influenza-like illness	WHO website
	Indonesia	H5N1	20	13	Influenza-like illness	WHO website
	Thailand	H5N1	5	2	Influenza-like illness	WHO website
	Vietnam	H5N1	61	19	Influenza-like illness	WHO website
2006	Azerbaijan	H5N1	8	5	Influenza-like illness	WHO website
	Cambodia	H5N1	2	2	Influenza-like illness	WHO website
	China	H5N1	13	8	Influenza-like illness	WHO website
2006	Djibouti	H5N1	1	0	Influenza-like illness	WHO website
	Egypt	H5N1	18	10	Influenza-like illness	WHO website
	Indonesia	H5N1	55	45	Influenza-like illness	WHO website
	Iraq	H5N1	3	2	Influenza-like illness	WHO website
	Thailand	H5N1	3	3	Influenza-like illness	WHO website
	Turkey	H5N1	12	4	Influenza-like illness	WHO website
September 2007	Cambodia	H5N1	1	1	Influenza-like illness	WHO website
	China	H5N1	2	1	Influenza-like illness	WHO website
	Egypt	H5N1	16	3	Influenza-like illness	WHO website
	Indonesia	H5N1	6	5	Influenza-like illness	WHO website
	Lao People's Democratic Republic	H5N1	2	2	Influenza-like illness	WHO website
	Nigeria	H5N1	1	1	Influenza-like illness	WHO website
	Vietnam	H5N1	7	4	Influenza-like illness	WHO website
	Total number of human deaths			179		
	Total number of human infections			417		

these viruses. For example, a serological survey of poultry workers in Hong Kong after the 1997 outbreak identified 10% seroprevalence of H5 antibodies, but without any known occurrence of clinical disease (Bridges et al. 2002). In relation to serological investigations in humans during the Dutch 2003 H7N7 epidemic, which also caused one human fatality and 83 confirmed cases of conjunctivitis, extensive seropositivity was reported (Fouchier et al. 2004). Despite this, no reassortant virus has emerged and it may well be that other, unknown factors limit the chances of a pandemic virus arising in this way.

The second mechanism by which the generation of a pandemic virus may occur is through progressive adaptation of a virus entirely of avian origin. Recent studies on the genome of the H1N1 "Spanish" influenza virus, which affected human beings at the beginning of the 20th century, have resulted in the speculation that this virus was entirely of avian origin and not generated by reassortment (Taubenberger 2005). Thus, a virus containing all eight segments of avian origin was able to establish itself in the human population and cause many more deaths than World War I. Sequencing of genes of this and other viruses that have infected humans directly from an avian source has highlighted mutations that are a result of or progressive adaptation to the human host.

Regardless of the mechanism by which the new human pandemic virus may be generated, it appears logical that AI virus circulation in animals should be reduced and that, above all, contacts at risk should be avoided. This is one of the most complex problems to be addressed in developing countries. The human cases that have occurred during the ongoing H5N1 epidemic have usually developed following contacts between villagers and rural chickens or fighting cocks. The nature and entity of these contacts are dependent on social and behavioural practices linked to food security or hobby activities. In addition, basic hygienic standards are rarely respected. The modification of these patterns appears to be inapplicable in the short term. Efforts should be concentrated on the reduction of viral shedding from rural poultry so that the amount of virus shed is insufficient to infect a human being. It is considered by many that vaccination interventions of rural poultry are currently the only means to achieve a reduction of virus load in the rural environment. However, if this is attempted without putting in place other important measures, including extension services to ensure vaccination is carried out correctly, proper surveillance of vaccinated birds such that infected birds can be detected and infections eradicated, it is unlikely that vaccination alone will have the desired effect and may well make the situation worse.

In order to control infection of rural poultry, the awareness of AI and of the risk it poses should increase. This implies the education of farmers and of poultry workers regarding the basic concepts of biosecurity, farming hygiene, prevention and notification procedures. Farmers should self-notify outbreaks rather than attempting to escape restrictions; they should also be trained in outbreak management practices, including recognition of the disease as well as the culling of infected birds and their appropriate disposal. In case of the implementation of a vaccination campaign, it is imperative that it is carried out using hygienic and appropriate logistic/management practices. Vaccine must be of high quality and administered to each group of birds with sterile syringes. The cold chain must be respected and vaccine bottles must be shaken vigorously prior to use, so that the quality of the product is maintained and efficacy is guaranteed.

In these conditions, field exposure of infected flocks can be rather difficult to assess. Laboratory diagnosis is not performed in an extensive manner when it comes to rural poultry, and vaccinated birds may not display any clinical signs but actively shed virus, thus perpetuating infection. Leaving unvaccinated sentinel birds in the flock appears to be the only pursuable system of detecting field exposure. The identification of sentinels in rural establishments could be achieved by leaving male birds in the flock unvaccinated.

1.8 Conclusion

The modified ecology and epidemiology of AI infections requires urgent progress in our knowledge of issues related to epidemiology, pathogenesis and control.

The Asian lineage H5N1 viruses have spread to three continents, each with completely different agricultural, ecological, social and economic backgrounds. This, in turn, has resulted in the establishment of diverse mechanisms by which the virus is perpetuated in a given area. The generation of these cycles is also influenced by the diversity and availability of hosts in that context. As the virus encounters new hosts—within and outside the *Class Aves*—it acquires mutations that may reflect pathogenetic advantages in one or more species.

The results of these two driving forces in the genetic and antigenic profile require the monitoring of viral strains and a close collaboration between the parties involved in crisis management. The monitoring effort should aim at the collection and characterisation of strains in order to identify genetic mutations and antigenic properties. Information should be collated and made available to the international scientific community, particularly medical virologists.

At least two AI subtypes, H5N1 and H9N2, both of which have zoonotic implications, are currently apparently endemic in vast areas of the world. It is impossible to predict whether either of them will represent the progenitor of the next human pandemic virus. Certainly, both of them are causing losses to the poultry industry and H5N1 is also responsible for the loss of human lives and the reduction of the livelihood of rural establishments.

The medical, veterinary and agricultural scientific communities are challenged with a virus that is moving in a tri-dimensional fashion, modifying itself as it adapts to different species and reassorting with other influenza viruses of avian and potentially mammalian origin as it infects new species. Significant collaborative and financial efforts carried out in a transparent scientific environment are required to generate data and ideas contributing to the eradication effort. Until extensive circulation of the virus is limited in the poultry reservoir, AI will continue to remain an issue for food security and a global threat for animal and human health. The veterinary community should take ownership of this responsibility, as it has the knowledge and understanding to offer sustainable solutions to the management of this infection.

References

Alexander DJ (2000) A review of avian influenza in different bird species. Vet Microbiol 74(1-2):3-13

Alexander DJ (2001) Ecology of avian influenza in domestic birds. Proceedings of the International Symposium on Emergence and Control of Zoonotic Ortho- and Paramyxovirus Diseases. Merieux Foundation, Veyer du Lac, France 25-34

Alexander DJ (2002) Report on avian influenza in the Eastern Hemisphere during 1997-2002. Avian Dis 47 (3 Suppl):792-797

Alexander DJ (2008) Avian influenza manual for diagnostic tests and vaccines for terrestrial animals, 6th edition. Chapter 2.7.12. World Organisation for Animal Health, Paris, France. http://www.oie.int/eng/normes/ mmanual/a_00002.htm

Alexander DJ (2007) Summary of avian influenza activity in Europe, Asia, Africa and Australasia 2002-2006. Avian Dis 51(1 Suppl):161-166

Alexander DJ, Allan WH, Parsons DG, Parsons G (1978) The pathogenicity of four avian influenza viruses for fowls, turkeys and ducks. Res Vet Sci 24(2):242-247

Alexander DJ, Parsons G, Manvell RJ (1986) Experimental assessment of the pathogenicity of eight avian influenza A viruses of H5 subtype for chickens, turkeys, ducks and quails. Avian Pathol 15:647-662

Banks J, Speidel E, Alexander DJ (1998) Characterisation of an avian influenza A virus isolated from a human--is an intermediate host necessary for the emergence of pandemic influenza viruses? Arch Virol 143(4):781-787

Banks J, Speidel EC, Harris PA, Alexander DJ (2000) Phylogenetic analysis of influenza A viruses of H9 haemagglutinin subtype. Avian Pathol 29:353-360

Banks J, Speidel EC, McCauley JW, Alexander DJ (2000) Phylogenetic analysis of H7 haemagglutinin subtype influenza A viruses. Arch Virol 145(5):1047-1058

Beare AS, Webster RG (1991) Replication of avian influenza viruses in humans. Arch Virol 119(1-2):37-42

Becker WB (1966) The isolation and classification of Tern virus: Influenza A - Tern South Africa - 1961. J Hyg (Lond) 64(3):309-320

Berg M, Englund L, Abusugra IA et al (1990) Close relationship between mink influenza (H10N4) and concomitantly circulating avian influenza viruses. Arch Virol 113(1-2):61-71

Bridges CB, Lim W, Hu-Primmer J et al (2002) Risk of influenza A (H5N1) infection among poultry workers, Hong Kong, 1997-1998. J Infect Dis 185(8):1005-1010

Butt KM, Smith GJ, Chen H et al (2005) Human infection with an avian H9N2 influenza A virus in Hong Kong in 2003. J Clin Microbiol 43(11):5760-5767

Callan RJ, Early G, Kida H, Hinshaw VS (1995) The appearance of H3 influenza viruses in seals. J Gen Virol 76(Pt 1):199-203

Campbell CH, Webster RG, Breese SS Jr (1970) Fowl plague virus from man. J Infect Dis 122(6):513-516

Campbell G (1998) Report of the Irish national reference laboratory for 1996 and 1997. Proceedings of the Joint Fourth Annual Meetings of the National Newcastle Disease and Avian Influenza Laboratories of Countries of the European Union, Brussels, 1997 p13

Capua I, Alexander DJ, (2004) Avian influenza: recent developments. Avian Pathol 33(4):393-404

Capua I, Marangon S (2007) The use of vaccination to combat multiple introductions of Notifiable Avian Influenza viruses of the H5 and H7 subtypes between 2000 and 2006 in Italy. Vaccine 25(27):4987-4995

Capua I, Marangon S, Dalla Pozza M et al (2003) Avian

influenza in Italy 1997-2001. Avian Dis 47(3 Suppl): 839-843

Capua I, Mutinelli F (2001) An atlas and text on avian influenza. Papi Editore pp 1-236

Capua I, Mutinelli F, Marangon S, Alexander DJ (2000) H7N1 Avian Influenza in Italy (1999-2000) in intensively reared chickens and turkeys. Avian Pathol 29:737-743

Chambers TM, Yamnikova S, Kawaoka Y et al (1989) Antigenic and molecular characterization of subtype H13 hemagglutinin of influenza virus. Virology 172(1):180-188

Chan PK (2002) Outbreak of avian influenza A (H5N1) virus infection in Hong Kong in 1997. Clin Infect Dis 34(2 Suppl):58–64

Chen H, Smith GJ, Zhang SY et al (2005) Avian flu: H5N1 virus outbreak in migratory waterfowl. Nature 436(7048):191-192

Choi YK, Nguyen TD, Ozaki H et al (2005) Studies of H5N1 influenza virus infection of pigs by using viruses isolated in Vietnam and Thailand in 2004. J Virol 79(16):10821-10825

Davison S, Eckroade RJ, Ziegler AF (2003) A review of the 1996-98 nonpathogenic H7N2 avian influenza outbreak in Pennsylvania. Avian Dis 47(3 Suppl): 823-827

EFSA (2005) Epidemiology report on avian influenza in a quarantine premises in Essex http://defra.gov.uk/animalh/diseases/notifiable/disease/ai/pdf/ai-epidemrep 111105.pdf

EFSA (2005) Animal health and welfare aspects of avian influenza. EFSA J 266:1-21

Ellis TM, Bousfield RB, Bissett LA et al (2004) Investigation of outbreaks of highly pathogenic H5N1 avian influenza in waterfowl and wild birds in Hong Kong in late 2002. Avian Pathol 33(5):492-505

Fioretti A, Menna LF, Calabria M (1998) The epidemiological situation of avian influenza in Italy during 1996-1997. Proceedings of the Joint Fourth Annual Meetings of the National Newcastle Disease and Avian Influenza Laboratories of Countries of the European Union, Brussels 1997, 17-22

Fouchier RA, Schneeberger PM, Rozendaal FW et al (1980) Avian influenza A virus (H7N7) associated with human conjunctivitis and a fatal case of acute respiratory distress syndrome. Proc Natl Acad Sci USA 101(5):1356-1361

Gething MJ, Bye J, Skehel J, Waterfield M (1980) Cloning and DNA sequence of double-stranded copies of haemagglutinin genes from H2 and H3 strains elucidates antigenic shift and drift in human influenza virus. Nature 287(5780):301-306

Glass SE, Naqi SA, Grumbles LC (1981) Isolation of avian influenza virus in Texas. Avian Dis 25(2):545-549

Guan Y, Shortridge KF, Krauss S et al (1996) Emergence of avian H1N1 influenza viruses in pigs in China. J Virol 70(11):8041-8046

Guo Y, Wang M, Kawaoka Y et al (1992) Characterization of a new avian-like influenza A virus from horses in China. Virology 188(1):245-255

Halvorson DA, Frame DD, Friendshuh AJ, Shaw DP (1998) Outbreaks of low pathogenicity avian influenza in USA. Proceedings of the 4th International Symposium on Avian Influenza, Athens, Georgia, US Animal Health Association 36-46

Halvorson DA, Karunakaran D, Senne D et al (1983) Epizootiology of avian influenza -- simultaneous monitoring of sentinel ducks and turkeys in Minnesota. Avian Dis 27(1):77-85

Halvorson DA, Kelleher CJ, Pomeroy BS et al (1987) Surveillance procedures for avian influenza. Proceedings of the Second International Symposium on Avian Influenza, University of Wisconsin, Madison 155-63

Halvorson DA, Kodillalli S, Laudert E et al (1992) Influenza in turkeys in turkey in the USA, 1987-1991. Proceedings of the 3rd International Symposium on Avian Influenza 33-42

Hinshaw VS, Bean WJ, Geraci J et al (1986) Characterization of two influenza A viruses from a pilot whale. J Virol 58(2):655-656

Hinshaw VS, Webster RG, Easterday BC, Bean WJ Jr (1981b) Replication of avian influenza A viruses in mammals. Infect Immun 34(2):354-361

Hinshaw VS, Webster RG, Rodriguez RJ (1981a) Influenza A viruses: combinations of hemagglutinin and neuraminidase subtypes isolated from animals and other sources. Arch Virol 67(3):191-201

Homme PJ, Easterday BC, Anderson DP (1970) Avian influenza virus infections. II. Experimental epizootiology of influenza A-turkey-Wisconsin-1966 virus in turkeys. Avian Dis 14(2):240-247

Humberd J, Guan Y, Webster RG (2006) Comparison of the replication of influenza A viruses in Chinese ring-necked pheasants and chukar partridges. J Virol 80(5):2151-2161

Irvine RM, Banks J, Londt BZ et al (2007) An outbreak of neglypathogenic avian influenza caused by an Asian lineage H5N1 virus in turkeys in Great Britain in January 2007. Vet Rec 161:100-101

Isoda N, Sakoda Y, Kishida N et al (2006) Pathogenicity of a highly pathogenic avian influenza virus, A/chicken/Yamaguchi/7/04 (H5N1) in different species of birds and mammals. Arch Virol 151(7):1267-1279

Johnson DC (1984) AI task force veterinarian offers practical suggestions. Broiler Indust 47:58-59

Karasin AI, Brown IH, Carman S, Olsen CW (2000) Isolation and characterization of H4N6 avian influenza viruses from pigs with pneumonia in Canada. J Virol 74(19):9322-9327

Karasin AI, West K, Carman S, Olsen CW (2004) Characterization of avian H3N3 and H1N1 influenza A viruses isolated from pigs in Canada. J Clin Microbiol 42(9):4349-4354

Kawaoka Y, Chambers TM, Sladen WL, Webster RG (1988) Is the gene pool of influenza viruses in shorebirds and gulls different from that in wild ducks? Virology 163(1):247-250

Kawaoka Y, Krauss S, Webster RG (1989) Avian-to-human transmission of the PB1 gene of influenza A viruses in the 1957 and 1968 pandemics. J Virol 63(11):4603-4608

Keawcharoen J, Oraveerakul K, Kuiken T et al (2004) Avian influenza H5N1 in tigers and leopards. Emerg Infect Dis 10(12):2189-2191

Kida H, Ito T, Yasuda J et al (1994) Potential for transmission of avian influenza viruses to pigs. J Gen Virol 75(Pt 9):2183-2188

King LJ (1984) How APHIS "war room" mobilized to fight AI. Broiler Indust 47:44-51

Klingeborn B, Englund L, Rott R et al (1985) An avian influenza A virus killing a mammalian species -- the mink. Brief Report. Arch Virol 86(3-4):347-351

Kuiken T, Rimmelzwaan G, van Riel D et al (2004) Avian H5N1 influenza in cats. Science 306(5694):241

Kurtz J, Manvell RJ, Banks J (1996) Avian influenza virus isolated from a woman with conjunctivitis. Lancet 348(9031):901-902

Lang G (1982) A review of influenza in Canadian domestic and wild birds. Proceedings of the First International Symposium on Avian influenza, Carter Composition Corporation, Richmond, USA 21-27

Lang G, Gagnon A, Geraci JR (1981) Isolation of an influenza A virus from seals. Arch Virol 68(3-4):189-195

Lei F, Tang S, Zhao D et al (2007) Characterization of H5N1 influenza viruses isolated from migratory birds in Qinghai province of China in 2006. Avian Dis 51(2):568-572

Li SQ, Orlich M, Rott R (1990) Generation of seal influenza virus variants pathogenic for chickens, because of hemagglutinin cleavage site changes. J Virol 64(7):3297-3303

Liu J, Xiao H, Lei F et al (2005) Highly pathogenic H5N1 influenza virus infection in migratory birds. Science 309(5738):1206

Londt BZ, Banks J, Alexander DJ (2007) Highly pathogenic avian influenza viruses with low virulence for chickens in in vivo tests. Avian Pathol 36(5):347-350

Lvov D (1978) Circulation of influenza viruses in natural biocoenosis. Viruses and Environment 351-380

Loeffen W, De Boer-Luitze E, Koch G (2003) Infection with avian influenza virus (H7N7) in Dutch pigs. Proceedings ESVV Congress St Malo France 50

Loeffen W, De Boer-Luitze E, Koch G (2004) Transmission of a highly pathogenic avian influenza virus to swine in the Netherlands Proceedings of the in-between congress of the International Society for Animal Hygiene 329-330

Ludwig S, Haustein A, Kaleta EF, Scholtissek C (1994) Recent influenza A (H1N1) infections of pigs and turkeys in northern Europe. Virology 202(1):281-286

Makarova NV, Ozaki H, Kida H et al (2003) Replication and transmission of influenza viruses in Japanese quail. Virology 310(1):8-15

Mo IP, Song CS, Kim KS, Rhee JC (1998) An occurrence of non-highly pathogenic avian influenza in Korea. Proceedings of the 4th International Symposium on Avian Influenza, Athens, Georgia (1997) US Animal Health Association 379-83

Mohan R, Saif YM, Erickson GA et al (1981) Serologic and epidemiologic evidence of infection in turkeys with an agent related to the swine influenza virus. Avian Dis 25(1):11-16

Monne I, Joannis TM, Fusaro A et al (2008) Reassortant avian influenza virus (H5N1) in poultry, Nigeria, 2007. Emerg Infect Dis 14(4):637-640. Available from http://www.cdc.gov/EID/content/14/4/637.htm

Narayan O, Lang G, Rouse BT (1969) A new influenza A virus infection in turkeys. V. Pathology of the experimental disease by strain turkey-Ontario 7732-66. Arch Gesamte Virusforsch 26(1):166-182

OIE (2006) Update on avian influenza in animals (type H5), April 07 2006, http://www.oie.int/downld/avian%20influenza/A_AI-Asia.htm

OIE (2007) Update on avian influenza in animals (typeH5), September 15 2007, http://www.oie.int/downld/avian%20influenza/A_AI-Asia.htm

Okazaki K, Yanagawa R, Kida H (1983) Contact infection of mink with 5 subtypes of avian influenza virus. Brief report. Arch Virol 77(2-4):265-269

Olsen B, Munster VJ, Wallensten A et al (2006) Global patterns of influenza a virus in wild birds. Science 312(5772):384-388

Pasick J, Handel K, Robinson J, Copps J et al (2005) Intersegmental recombination between the haemagglutinin and matrix genes was responsible for the emergence of a highly pathogenic H7N3 avian influenza virus in British Columbia. J Gen Virol 86(Pt 3):727-731

Peiris JS, Guan Y, Markwell D et al (2001) Cocirculation of avian H9N2 and contemporary "human" H3N2 influenza A viruses in pigs in southeastern China: potential for genetic reassortment? J Virol 75(20):9679-9686

Peiris M, Yuen KY, Leung CW et al (1999) Human infection with influenza H9N2. Lancet 354(9182):916-917

Pensaert M, Ottis K, Vandeputte J et al (1981) Evidence for the natural transmission of influenza A virus from wild ducks to swine and its potential importance for man. Bull World Health Org 59(1):75-78

Perdue ML, Crawford JM, Garcia M et al (1998) Occurrence and possible mechanisms of cleavage site insertions in the avian influenza hemagglutinin gene. Proceedings of the 4th International Symposium on Avian Influenza, Athens, Georgia. US Animal Health Association, 182-193

Perez DR, Webby RJ, Hoffmann E, Webster RG (2003) Land-based birds as potential disseminators of avian mammalian reassortant influenza A viruses. Avian Dis 47(3 Suppl):1114-1117

Pomeroy BS (1982) Avian influenza in the United States (1964-1980). Proceedings of the First International Symposium on Avian Influenza 13-17

Pomeroy BS (1987) Avian influenza - Avian influenza in turkeys in the USA. Proceedings of the Second International Symposium on Avian Influenza, University of Wisconsin, Madison. 14-21

Puzelli S, Di Trani L, Fabiani C et al (2005) Serological analysis of serum samples from humans exposed to avian H7 influenza viruses in Italy between 1999 and 2003. J Infect Dis 192(8):1318-1322

Röhm C, Horimoto T, Kawaoka Y et al (1995) Do hemagglutinin genes of highly pathogenic avian influenza viruses constitute unique phylogenetic lineages? Virology 209(2):664-670

Salzberg SL, Kingsford C, Cattoli G et al (2007) Genome analysis linking recent European and African influenza (H5N1) viruses. Emerg Infect Dis 13(5):713-718

Scholtissek C, Koennecke I, Rott R (1978) Host range recombinants of fowl plague (influenza A) virus. Virology 91(1):79-85

Senne DA (2007) Avian influenza in North and South America, 2002-2005. Avian Dis 51(1 Suppl):167-173

Senne DA (2003) Avian influenza in the Western Hemisphere including the Pacific Islands and Australia. Avian Dis 47(3 Suppl):798-805

Senne DA, Panigrahy B, Kawaoka Y, Pearson JE et al (1996) Survey of the haemagglutinin (HA) cleavage site sequence of H5 and H7 avian influenza viruses: amino acid sequence at the HA cleavage site as a marker of pathogenicity potential. Avian Dis 40(2):425-437

Senne DA, Suarez DL, Stallnecht DE, Pedersen JC et al (2006) Ecology and epidemiology of avian influenza in North and South America. Dev Biol 124:37-44

Sharp GB, Kawaoka Y, Wright SM et al (1993) Wild ducks are the reservoir for only a limited number of influenza A subtypes. Epidemiol Infect 110(1):161-176

Shortridge KF, Zhou NN, Guan Y et al (1998) Characterization of avian H5N1 influenza viruses from poultry in Hong Kong. Virology 252(2):331-342

Sims LD, Domenech J, Benigno C et al (2005) Origin and evolution of highly pathogenic H5N1 avian influenza in Asia. Vet Rec 157(6):159-164

Songserm T, Amonsin A, Jam-on R et al (2006) Fatal avian influenza A H5N1 in a dog. Emerg Infect Dis 12(11):1744-1747

Songserm T, Amonsin A, Jam-on R et al (2006) Avian influenza H5N1 in naturally infected domestic cat. Emerg Infect Dis 12(4):681-683

Songserm T, Jam-on R, Sae-Heng N et al (2006) Domestic ducks and H5N1 influenza epidemic, Thailand. Emerg Inf Dis 12(4):575-581

Stallknecht DE (1998) Ecology and epidemiology of avian influenza viruses in wild birds populations. Proceedings of the Fourth International Symposium on Avian Influenza, 61-69

Stallknecht DE, Shane SM (1988) Host range of avian influenza virus in free-living birds. Vet Res Commun 12(2-3):125-141

Stieneke-Gröber A, Vey M, Angliker H, Shaw E et al (1992) Influenza virus hemagglutinin with multibasic cleavage site is activated by furin, a subtilisin-like endoprotease. EMBO J 11(7):2407-2414

Suarez DL, Senne DA, Banks J, Brown IH et al (2004) Recombination resulting in virulence shift in avian influenza outbreak, Chile. Emerg Infect Dis 10(4):693-699

Swayne DE, Alexander DJ (1994) Confirmation of nephrotropism and nephropathogenicity of three low-pathogenic chicken-origin influenza viruses for chickens. Avian Pathol 23(2):345-352

Taubenberger JK (2005) The virulence of the 1918 pandemic influenza virus: unraveling the enigma. Arch Virol Suppl (19):101-115

Taylor HR, Turner AJ (1977) A case report of fowl plague keratoconjunctivitis. Br J Ophthalmol 61(2):86-88

Thanawongnuwech R, Amonsin A, Tantilertcharoen R et al (2005) Probable tiger-to-tiger transmission of avian influenza H5N1. Emerg Infect Dis 11(5):699-701

Tsukamoto K, Imada T, Tanimura N et al (2007) Impact of different husbandry conditions on contact and airborne transmission of H5N1 highly pathogenic avian influenza virus to chickens. Avian Dis 51(1):129-132

Tûmová B (1980) Equine influenza--a segment in influenza virus ecology. Comp Immunol Microbiol Infect Dis 3(1-2):45-59

Utterback W (1984a) Update on avian influenza through February 21, 1984 in Pennsylvania and Virginia. Proceedings of the 33rd Western Poultry Disease Conference, 4-7

Van Borm S, Thomas I, Hanquet G et al (2005) Highly Pathogenic H5N1 influenza virus in smuggled Thai eagles, Belgium. Emerg Infect Dis 11(5):702-705

Van Reeth K (2007) Avian and swine influenza viruses: our current understanding of the zoonotic risk. Vet Res 38(2):243-260

Vey M, Orlich M, Adler S et al (1992) Hemagglutinin activation of pathogenic avian influenza viruses of serotype H7 requires the protease recognition motif R-X-K/R-R. Virology 188(1):408-413

Webster RG, Bean WJ, Gorman OT et al (1992) Evolution and ecology of influenza A viruses. Microbiol Rev 56(1):152-179

Webster RG, Hinshaw VS, Bean WJ et al (1981) Characterization of an influenza A virus from seals. Virology 113(2):712-724

Wells RJH (1963) An outbreak of fowl plague in turkeys. Vet Rec 75:783-786

Werner O (1998) Avian influenza – Situation in Germany 1995-1997. Proceedings of the Joint Fourth Annual Meetings of the National Newcastle Disease and Avian Influenza Laboratories of Countries of the European Union, Brussels, 1997, 9-10

Werner O (1999) Avian influenza – Situation in Germany 1997/1998. Proceedings of the Joint Fifth Annual Meetings of the National Newcastle Disease and Avian Influenza Laboratories of Countries of the European Union, Vienna, 1998, 10-11

Westbury HA, Turner AJ, Amon C (1981) Transmissibility of two avian influenza A viruses (H7N7) between chicks. Avian Pathol 10:481-487

Westbury HA, Turner AJ, Kovesdy L (1979) The pathogenicity of three Australian fowl plague viruses for chickens, turkeys and ducks. Vet Microbiol 4:223-234

Wood GW, Banks J, Brown IH et al (1997) The nucleotide sequence of the HA1 of the haemagglutinin of an HI avian influenza virus isolate from turkeys in Germany provides additional evidence suggesting recent trasmission from pigs. Avian Pathol 26(2):347-355

Wood GW, McCauley JW, Bashiruddin JB, Alexander DJ (1993) Deduced amino acid sequences at the haemagglutinin cleavage site of avian influenza A viruses of H5 and H7 subtypes. Arch Virol 130(1-2):209-217

Wright SM, Kawaoka Y, Sharp GB et al (1992) Interspecies transmission and reassortment of influenza A viruses in pigs and turkeys in the United States. Am J Epidemiol 136(4):488-497

Xu X, Subbarao K, Cox, NJ, Guo Y (1999) Genetic characterization of the pathogenic influenza A/Goose/Guandong/1/96 (H5N1) virus: similarity of its hemagglutinin gene to those of H5N1 viruses from the 1997 outbreaks in Hong Kong. Virology 261(1):15-19

Yingst SL, Saad MD, Felt SA (2006) Qinghai-like H5N1 from domestic cats, northern Iraq. Emerg Infect Dis 12(8):1295-1297

Yuen KY, Chan PK, Peiris M et al (1998) Clinical features and rapid viral diagnosis of human disease associated with avian influenza A H5N1 virus. Lancet 351(9101):467-471

Ecology and Epidemiology of Newcastle Disease

2

Dennis J. Alexander

2.1 Introduction

The first outbreaks of the severe disease of poultry known as Newcastle disease (ND) occurred in 1926, in Java, Indonesia (Kraneveld 1926), and in Newcastle-upon-Tyne, England (Doyle 1927). The name "Newcastle disease" was coined by Doyle as a temporary measure because he wished to avoid a descriptive name that might be confused with other diseases (Doyle 1935). The name has, however, continued to be used, although when referring to ND virus (NDV), the synonym "avian paramyxovirus type 1" (APMV-1) is now often employed. Sometimes APMV-1 has been used to describe ND strains of low virulence, to avoid terming them ND viruses, as the definitions used by the World Organisation for Animal Health (Alexander 2008) and other international agencies reserve ND for virulent viruses.

Whether the outbreaks of 1926 marked the emergence of ND has been the subject of some discussion, as there are earlier reports of similar disease outbreaks in Central Europe before this date (Halasz 1912). Macpherson (1956), in reviewing the death of all the chickens in the Western Isles of Scotland in 1896, considered it probable that the cause was ND. It is possible, therefore, that ND did occur in poultry before 1926, but its recognition as a specifically defined disease of viral aetiology dates from the outbreaks during that year in Newcastle-upon-Tyne.

Later, it became clear that other, less severe infections were caused by viruses almost identical to the original virus. In the United States, a relatively mild respiratory disease, often with nervous signs, was first reported in the 1930s and subsequently termed "pneumoencephalitis" (Beach 1942). It was shown to be due to a virus indistinguishable from NDV in serological tests (Beach 1944). Since then, numerous isolations of viruses that produce an extremely mild disease or no evidence of disease in chickens have been made around the world, and it is now accepted that pools of such viruses are perpetuated in waterfowl and other wild birds.

2.2 Aetiology

The virus order Mononegavirales (i.e. the single-stranded, nonsegmented, negative-sense RNA viruses showing helical capsid symmetry) is formed from the virus families *Paramyxoviridae*, *Filoviridae* and *Rhabdoviridae*. The family *Paramyxoviridae* is divided into two subfamilies *Paramyxovirinae* and *Pneumovirinae* (Lamb et al. 2005). The subfamily Paramyxovirinae has five genera: *Rubulavirus,* which includes the mumps virus, mammalian para-influenza 2 and 4; *Respirovirus* containing mammalian para-influenza viruses 1 and 3; *Morbillivirus,* measles, distemper and rinderpest; *Henipavirus*, formed from the Nipah and Hendra viruses; and *Avulavirus*, formed from NDV and other avian paramyxoviruses (Lamb et al. 2005).

Nine serogroups of avian paramyxoviruses have been recognised: APMV-1 to APMV-9 (Alexander 1988a). Of these, NDV (APMV-1) remains the most important pathogen for poultry, but APMV-2, APMV-3, APMV-6 and APMV-7 are known to cause disease in poultry. The nomenclature used for isolates of influenza A virus has been adopted for avian paramyxoviruses so that an isolate is named by: (1) serotype, (2) species or type of bird from which it was isolated, (3) geographical location of isolation, (4) reference number or name and (5) year of isolation.

Antigenic variation of ND viruses (APMV-1) detectable by conventional haemagglutination inhibition (HI) tests has been reported, although such reports are rare and represent relatively minor variations (Arias-Ibarrondo et al. 1978; Hannoun 1977; Alexander et

I. Capua, D.J. Alexander (eds.) *Avian Influenza and Newcastle Disease*,

al. 1984). One of the most noted variations of this kind has been the virus responsible for the panzootic in racing pigeons. This NDV, often referred to as pigeon APMV-1 (PPMV-1), was demonstrably different from standard strains in haemagglutination inhibition tests, but not sufficiently different antigenically that conventional ND vaccines were not protective (Alexander and Parsons 1986). Antigenic variations between ND strains have been detected by monoclonal antibodies (mAbs) and have been used as an epidemiological tool (Alexander et al. 1997). Although use of mAb panels has shown that viruses grouped by their ability to react with the same mAbs share biological and epidemiological properties, this approach to understanding the epidemiology of ND has been largely replaced by phylogenetic analysis.

Genetic techniques have become established in the diagnosis of ND and as an epidemiological tool for distinguishing between virus strains (Aldous and Alexander 2001). Herczeg et al. (1999, 2001); Lomniczi et al. (1998) concluded from their phylogenetic analyses of NDV isolates that there were eight genetic lineages (I–VIII) and several sublineages within them. Aldous et al. (2003), in a study of 338 isolates of NDV representing a range of viruses of different temporal, geographical and host origins, concluded that the isolates divided into six broadly distinct groups (lineages 1–6). Lineages 3 and 4 were further subdivided into four sublineages (a–d) and lineage 5 into five sublineages (a–e). Essentially, lineages 1, 2, 4 and 5 correspond to the earlier defined lineages I, II, VI and VII, with comparable sublineages but the geno-groupings III, IV, V, VIII correspond to the sublineages 3a–3d. Lineage 6 represents a new geno-group.

Although the NDV isolates placed in geno-groups 1–5 (or I–VIII) are genetically quite close, viruses that were placed in geno-group 6 by Aldous et al. (2003), and later class I by Czeglédi et al. (2006), are very different from all the other NDV isolates, i.e. the class II viruses (Czeglédi et al. 2006). This has caused problems in molecular diagnosis, particularly as different primers are necessary for their detection in RT-PCR tests.

2.3 Host Range

Following a review of the available literature, Kaleta and Baldauf (1988) concluded that, in addition to the domestic avian species, natural or experimental infection with NDV has been demonstrated in at least 241 species from 27 of the 50 Orders of birds. It is highly probable that all bird species are susceptible to infection, but the outcome of infection in terms of disease varies considerably with different species.

2.3.1 Domestic Poultry

Virulent NDV strains have been isolated from all types of commercially reared poultry, ranging from pigeons to ostriches. The disease signs seen in different poultry infected with virulent NDV may show considerable variation. Ducks, for example, may not show clinical signs, while in other species the disease may be milder than in chickens and cause problems in initial diagnosis, e.g. in pheasants (Aldous et al. 2007). Ostriches may also cause problems in the initial suspicion of ND since, while they have been reported to show typical nervous signs, there is some difference in the severity of disease between young and adult birds (Alexander 2000).

Marginal domestic poultry may also play a significant role in the epidemiology of ND. For example, fighting cocks were involved in outbreaks of ND in the United States on several occasions. The most notable outbreak occurred in southern California in 2002–2003 (Kinde et al. 2003), where the widespread presence of ND in fighting cocks and the mobility and value of such birds not only posed considerable control problems but resulted in spread to 21 commercial table-egg farms and the slaughter of 3 million birds. The highest risk factors for infected commercial flocks were the farm employees and proximity to infected backyard game fowl.

2.3.2 Wild Birds

Isolates of NDV have been obtained frequently from wild birds, especially migratory feral waterfowl and other aquatic birds. Most of these isolates have been of low virulence for chickens and similar to viruses of the "asymptomatic enteric" pathotype.

Occasionally, virulent viruses have been detected in wild birds, but usually these were in birds found dead near infected poultry. The most significant outbreaks of NDV in feral birds have been those reported in double-crested cormorants (*Phalacrocorax auritus*) in North America since the 1990s. These outbreaks began in 1990 in Canada, specifically, in Alberta, Saskatchewan and Manitoba (Wobeser et al. 1993). The disease re-appeared in 1992 in cormorants in mid-western Canada, around the Great Lakes and northern Midwest USA, in the latter case spreading to domestic turkeys (Mixson and Pearson 1992; Heckert 1993). Disease in double-crested cormorants was observed again in Canada in 1995 and in California in 1997; in both instances, NDV was isolated from dead birds (Kuiken 1998).

Antigenic and genetic analyses of the viruses isolated from the cormorants suggested that these viruses were very closely related despite the geographical separation of the hosts. Since these outbreaks covered birds that would follow different migratory routes it seems most probable that initial infection occurred at a mutual wintering area in the Southern USA or Central America. Allison et al. (2005) were able to isolate similar virulent NDV from double-crested cormorants over-wintering in the Florida Keys.

There had been earlier reports of ND in cormorants and related species in the late 1940s in Scotland (Blaxland 1951) and in 1975 in Quebec (Cleary 1977). It is therefore possible that cormorants represent an occasional or even continual reservoir of virulent NDV.

Interestingly, wild birds have been implicated in the introduction of virulent NDV into poultry in a number of outbreaks over the last 10 years. For example, it was concluded that the virus responsible for the outbreaks of ND in the UK in 1997 (Alexander et al. 1999) had most likely been introduced by migratory wild birds. The virus responsible for the outbreaks in free-living pheasants in Denmark in 1996 was closely related, as were isolates from a goosander in Finland in 1996 and, perhaps significantly, a cormorant from Denmark in 2001 (Jørgensen al. 1999; Alexander et al. 1999; P. Jørgensen, personal communication). Re-emergence of a genetically very closely related virus in pheasants in Great Britain and France in 2005 and the close proximity of the French (Loire Atlantique) farm to a lake led to the speculation that this virus may be established in some species of wild birds in Europe (Aldous et al. 2007).

2.3.3 Caged "Pet Birds"

Virulent NDV isolates have often been obtained from captive caged birds (Senne et al. 1983). Kaleta and Baldauf (1988) thought it unlikely that infections of recently imported caged birds resulted from enzootic infections in feral birds in the countries of origin. They considered that the infections more probably originated at holding stations before export, either as a result of enzootic NDV at those stations or of spread from nearby poultry, such as backyard chicken flocks. Panigrahy et al. (1993) described outbreaks of severe ND in pet birds in six states in the USA in 1991. Illegal importations were assumed to be responsible for the introductions of the virus.

One important consideration for psittacines has been the demonstration that infected birds have been shown to excrete virulent NDV intermittently for extremely long periods, in some cases for more than a year (Erickson et al. 1977), which further emphasises the role these birds may have in the introduction of NDV to a country or area.

2.3.4 Racing and Show Pigeons

In the late 1970s, an NDV strain showing some antigenic differences from classical strains appeared in pigeons. This strain, PPMV-1, probably arose in the Middle East. In Europe, it was first reported in racing pigeons in Italy in 1981 (Biancifiori and Fioroni 1983) and subsequently produced a true panzootic, spreading in racing and show pigeons throughout the world (Aldous et al. 2004). The disease in pigeons has been recognised for over 25 years but still seems to remain enzootic in racing pigeons in many countries, with regular spread to wild pigeons and doves and a continuing threat to poultry.

2.4 Molecular Basis of Viral Virulence

An understanding of the molecular basis that controls the virulence of NDV strains (Rott and Klenk 1988) has meant that it is now possible, using nucleotide sequencing techniques, to assess whether or not an isolate has the genetic makeup to be highly pathogenic for poultry (Collins et al. 1993). The viral F protein brings about fusion between the viral membrane and the cell membrane so that the viral genome enters the cell and replication can begin. The F protein is therefore essential for replication. However, during replication, NDV particles are produced with a precursor glycoprotein, F0, that has to be cleaved to F1 and F2 polypeptides, which remain bound by disulphide bonds, for the virus particles to be infectious. This post-translational cleavage is mediated by host cell proteases.

The cleavability of the F0 molecule has been shown to be related directly to the virulence of the viruses in vivo. Numerous studies have confirmed the presence of multiple basic amino acids at the F0 cleavage site in virulent viruses. Usually the sequence is ^{113}RQK/RR*F^{117} in virulent viruses, but most have a basic amino acid at position 112 as well. In contrast, viruses of low virulence usually have the sequence ^{113}K/RQG/ER*L^{117}. Thus, there appears to be the requirement of a basic amino acid at residue 113, a pair of basic amino acids at 115 and 116 plus a phenylalanine at residue 117 if the virus is to be virulent for chickens. The presence of these basic amino acids at these positions means that cleavage can be effected by a protease or proteases present in a wide range of host tissues and organs; but for lentogenic viruses, cleavage can occur only with proteases recognising a single arginine, i.e. trypsin-like enzymes. Therefore, in host cells the replication of lentogenic viruses is restricted to areas with trypsin-like enzymes, such as the respiratory and intestinal tracts, whereas virulent viruses can replicate and cause damage in a range of tissues and organs, resulting in a fatal systemic infection.

That the virulence of NDV strains is governed by the F0 cleavage site is sufficiently accepted that it has been incorporated into the definition of ND adopted by the World Organisation for Animal Health (OIE):

"Newcastle disease is defined as an infection of birds caused by a virus of avian paramyxovirus serotype 1 (APMV-1) that meets one of the following criteria for virulence:

a) The virus has an intracerebral pathogenicity index (ICPI) in day-old chicks (Gallus gallus) *of 0.7 or greater.*

or

b) Multiple basic amino acids have been demonstrated in the virus (either directly or by deduction) at the C-terminus of the F2 protein and phenylalanine at residue 117, which is the N-terminus of the F1 protein. The term 'multiple basic amino acids' refers to at least three arginine or lysine residues between residues 113 and 116. Failure to demonstrate the characteristic pattern of amino acid residues as described above would require characterisation of the isolated virus by an ICPI test."

In this definition, amino acid residues are numbered from the N-terminus of the amino acid sequence deduced from the nucleotide sequence of the F0 gene, 113–116 corresponds to residues -4 to -1 from the cleavage site.' (Alexander 2008).

Various studies using cDNA clones of NDV and reverse genetics techniques have been undertaken to determine the precise minimum amino acid motif at the F0 cleavage site to confer virulence (Peeters et al. 1999). De Leeuw et al. (2003) generated a range of viruses with substituted amino acids at the F0 cleavage site. They concluded that virulence required F at position 117, R at 116, K or R at 115 and R not K at 113. Interestingly, all their generated mutants reverted to the virulent motifs ^{112}RRQRR*F^{117} or ^{112}RRQKR*F^{117} after a single passage in chicks.

Although it appears that the amino acid sequence of the F0 protein cleavage site is the primary influence on real or potential virulence of NDVs, it should be borne in mind that other factors associated with other virus genes and proteins may cause variations in virulence. For example, using reverse genetic techniques it has been demonstrated that the HN protein may influence virulence (Huang et al. 2004; Römer-Oberdörfer et al. 2006). Similarly, the V protein has been shown to inhibit apoptosis in infected cells and its absence also may affect virulence (Mebatsion et al. 2001).

2.5 Transmission

It is reasonable to conclude for NDVs that infection can take place by virus inhalation, ingestion (Alexander 1988b) or contact with mucous membranes, especially the conjunctiva. Spread from one bird to another therefore depends on the availability of the virus from the infected bird in an infectious form. Excretion of virus is dependent on the organs in which the virus multiplies and, as discussed above, this may vary with viral pathotype. Birds showing respiratory disease presumably shed virus in aerosols of mucus that may be inhaled by or contact susceptible birds. Viruses that are mainly restricted to intestinal replication may be transferred by ingestion of contaminated faeces, either directly or in contaminated food or water, or by the production of small infective particles produced from dried faeces that may be inhaled or impinge on mucous membranes. The method of virus transmission probably depends on many environmental factors that may drastically affect the rate of spread. Viruses transmitted by the respiratory route in a community of closely situated birds, such as in an intensive broiler house, may spread with alarming rapidity. Viruses excreted in the faeces and transmitted chiefly by the oral/faecal route may spread extremely slowly, especially if birds are not in direct contact, e.g. in caged layers.

The significance of vertical transmission of NDVs, especially virulent viruses, which usually cause cessation of egg-laying in susceptible diseased birds, is not clear. There have been some reports of isolation of vaccinal virus from eggs laid by infected birds (e.g. Pospisil et al. 1991), and in one significant report Capua et al. (1993) were able to isolate virulent NDV from cloacal swabs taken from birds with high antibody titres to NDV and from eggs laid by those birds as well as from the hatched progeny.

2.6 Spread

Several reviews have addressed the way in which NDV may be introduced into a country or area and then subsequently spread from flock to flock (Lancaster 1966; Lancaster and Alexander 1975; Alexander 1988b).

As discussed above, pools of NDV, usually of low virulence for poultry, are maintained in wild bird populations and primary introduction in poultry populations may occur by direct or indirect contact with wild birds. There is good evidence from analyses of viruses isolated in Ireland in 1990 and during the outbreaks of ND in Australia beginning in 1998 that, on rare occasions, viruses of low virulence may mutate to high virulence (Alexander 2001; Westbury 2001). Virulent NDV has also been generated experimentally from low-virulence virus by passage in chickens (Shengqing et al. 2002). Also, as discussed above, virulent NDV may be present in wild birds and other sectors; primary introduction may come from contact with them.

Once in the poultry sector, secondary spread has been attributed to a number of different methods such as: (1) movement of live birds; (2) contact with other animals; (3) movement of people and equipment; (4) movement of poultry products; (5) airborne spread; (6) contaminated poultry feed; (7) contaminated water and (8) vaccines.

While some of these are self-explanatory, others require more careful examination or are less obvious. For example, it is clear that susceptible infected birds could be moved and therefore spread ND during the incubation period of the disease. What may be of greater importance is that clinically normal, vaccinated birds have been shown to excrete virulent virus following challenge (Alexander et al. 1999; Guittet et al. 1993; Parede and Young 1990) and thus represent a serious threat in terms of overt disease to unvaccinated birds that may come in contact with them either directly, e.g. by trade in birds, especially for backyard flocks, or indirectly.

The role of airborne spread of NDV also requires some consideration. In the past, spread of the virus in the air had been considered an important route in some outbreaks (Dawson 1973), but of no importance in others (Utterback and Schwartz 1973) even involving the same virus. Hugh-Jones et al. (1973) attempted to assess the survival of airborne virus and were able to detect virus at 64 m downwind, albeit in very low titres, in very large amounts of air sampled, but not 165 m downwind of an infected premises. These authors stressed the importance of environmental conditions, particularly relative humidity, on the likelihood of airborne spread. In recent years, airborne spread has not been an issue in reported outbreaks and there has nearly always been an alternative, more likely cause, particularly the movement of poultry and the agency of humans.

2.7 Distribution

The widespread use of NDV vaccines in commercial poultry throughout the world makes the true geographical distribution of ND difficult to assess. It is usually considered that virulent NDV is either enzootic or a cause of regular epizootics in poultry throughout most of Africa, Asia, Central America and parts of South America. In more developed areas, such as Western Europe, sporadic epizootics occur on a fairly regular basis despite the widespread use of vaccination. The OIE (2007) lists only five countries where the disease has never occurred (French Guiana, Guyana, New Caledonia, Samoa and Vanuatu), 58 countries with "demonstrated clinical disease" between July 2005 and June 2007, and a further 14 countries with "unresolved disease events".

2.8 Human Health

The first report in which NDV was described to be a human pathogen was published by Burnet, in 1943. In a review of ND as a zoonosis, (Chang 1981) recorded 35 published reports of NDV infections of humans between 1948 and 1971. Since that time, there have been few additional publications, which probably reflects the lack of serious, lasting effects resulting from such infections and the fact that they are commonplace.

The most frequently reported and best substantiated clinical signs in human infections have been eye infections, usually consisting of unilateral or bilateral reddening, excessive lachrymation, oedema of the eyelids, conjunctivitis and subconjunctival haemorrhage (Chang 1981). Although the effect on the eye may be quite severe, infections are usually transient, lasting no more than a day or two, and the cornea is not affected. Reports of other clinical symptoms in humans infected with NDV are less well substantiated, but occasionally a more generalised infection resulting in chills, headaches and fever, with or without conjunctivitis, has been reported (Chang 1981).

Human infections with NDV have usually resulted from direct contact with the virus, infected birds or carcases of diseased birds. There have been no reports of human to human spread. The types of people known to have been infected with NDV include: laboratory workers (usually as a result of accidental splashing of infective material into the eye), veterinarians in diagnostic laboratories (presumably as a result of contact with infective material during postmortem examinations), workers in broiler processing plants and vaccination crews, especially when live vaccines are given as aerosols or fine dust. Pedersden et al. (1990) reported significantly higher antibody titres to NDV in people who had known associations with poultry.

2.9 Conclusion

Newcastle disease remains enzootic in poultry or other avian sectors, such as racing pigeons, in many areas of the world and thus represents a constant threat to most birds reared domestically. Every commercial flock of poultry reared is influenced in some way by measures aimed at controlling ND and spread of the virus. A large majority of the countries rearing poultry commercially rely on vaccination to keep ND under control, but the disease nevertheless represents a major limiting factor for increasing poultry production in many countries.

The greatest impact of ND may well be on village or backyard chicken production. In developing countries throughout Asia, Africa, Central America and some parts of South America, the village chicken is an extremely important asset in that it represents a significant source of protein in the form of eggs and meat. However, ND is frequently responsible for devastating losses in village poultry. Social and financial restraints mean that the control of ND in village chickens in developing countries is extremely difficult, if not impossible. This situation impinges on the further development of commercial poultry production and the establishment of trade links.

References

Aldous EW, Alexander DJ (2001) Technical review: detection and differentiation of Newcastle disease virus (avian paramyxovirus type 1). Avian Pathol 30:117–128

Aldous EW, Fuller CM, Mynn JK, Alexander DJ (2004) A molecular epidemiological investigation of isolates of the variant avian paramyxovirus type 1 virus (PPMV-1) responsible for the 1978 to present panzootic in pigeons. Avian Pathol 33(2):258-269

Aldous EW, Manvell RJ, Cox WJ et al (2007) Outbreak of Newcastle disease in pheasants (Phasianus colchicus) in south-east England in July 2005. Vet Rec 160(14):482-484

Aldous EW, Mynn JK, Banks J, Alexander DJ (2003) A molecular epidemiological study of avian paramyxovirus type 1 (Newcastle disease virus) isolates by phylogenetic analysis of a partial nucleotide sequence of the fusion protein gene. Avian Pathol 32(3):239–356

Alexander DJ (1988a) Newcastle disease virus-An avian paramyxovirus. In: DJ Alexander (ed) Newcastle Disease Kluwer Academic, Boston, MA, pp11-22

Alexander DJ (1988b) Newcastle disease: Methods of spread. In: DJ Alexander (ed) Newcastle disease. Kluwer Academic, Boston, MA, pp 256-272

Alexander DJ (2000) Newcastle disease in ostriches (Struthio camelus) – A review. Avian Pathol 29:95-100

Alexander DJ (2001) Gordon Memorial Lecture. Newcastle disease. Br Poult Sci 42(1):5–22

Alexander DJ (2008) Newcastle disease World Organisation for Animal Health Manual of Diagnostic Tests and Vaccines for Terrestrial Animals, 6th ed. Chapter 2.3.14. OIE, Paris, pp 576-589

Alexander DJ, Banks J, Collins MS et al (1999) Antigenic and genetic characterisation of Newcastle disease viruses isolated from outbreaks in domestic fowl and turkeys in Great Britain during 1997. Vet Rec 145(15):417-421

Alexander DJ, Manvell RJ, Banks J et al (1999) Experimental assessment of the pathogenicity of the Newcastle disease viruses from outbreaks in Great Britain in 1997 for chickens and turkeys and the protection afforded by vaccination. Avian Pathol 28:501-512

Alexander DJ, Manvell RJ, Lowings JP et al (1997) Antigenic diversity and similarities detected in avian paramyxovirus type 1 (Newcastle disease virus) isolates using monoclonal antibodies. Avian Pathol 26(2):399-418

Alexander DJ, Parsons G (1986) Protection of chickens against challenge with the variant virus responsible for Newcastle disease in 1984 by conventional vaccination. Vet Rec 118(7):176-177

Alexander DJ, Russell PH, Collins MS (1984) Paramyxovirus type 1 infections of racing pigeons: 1 characterisation of isolated viruses. Vet Rec 114(18):444-446

Allan WH, Lancaster JE, Toth B (1978) Newcastle disease vaccines—their production and use FAO Animal Production Series No 10 FAO, Rome, 163pp

Allison AB, Gottdenker NL, Stallknecht DE (2005) Wintering of neurotropic velogenic Newcastle disease virus and West Nile virus in double-crested cormorants (Phalacrocorax auritus) from the Florida Keys. Avian Dis 49(2):292–297

Arias-Ibarrondo J, Mikami T, Yamamoto H et al (1978) Studies on a paramyxovirus isolated from Japanese sparrow-hawks (Accipiter virgatus gularis). I. Isolation and characterization of the virus. Nippon Juigaku Zasshi 40:315-323

Beach JR (1942) Avian pneumoencephalitis. Proceedings of the Annual Meeting of the US Livestock Sanitary Association 46:203-223

Beach JR (1944) The neutralization in vitro of avian pneumoencephalitis virus by Newcastle disease immune serum. Science 100(2599):361-362

Beard CW, Hanson RP (1984) Newcastle disease. In: Hofstad MS, Barnes HJ, Calnek BW et al (eds) Diseases of poultry 8th ed. Iowa State University Press, Ames, pp 452-470

Biancifiori F, Fioroni A (1983) An occurrence of Newcastle disease in pigeons: virological and serological studies on the isolates. Comp Immunol Microbiol Infect Dis 6(3):247-252

Blaxland JD (1951) Newcastle disease in shags and cormorants and its significance as a factor in the spread of this disease among domestic poultry. Vet Rec 63:731-733

Burnet FM (1943) Human infection with the virus of Newcastle disease of fowl. Med J Aust 2:313–314

Capua I, Scacchia M, Toscani T, Caporale V (1993) Unexpected isolation of virulent Newcastle disease virus from commercial embryonated fowls' eggs. Zentralbl Veterinarmed B 40(9-10):609-612

Chang PW (1981) Newcastle disease. In: Beran GW (ed) CRC handbook series in zoonoses section B: Viral zoonoses, volume II. CRC, Baton Raton pp261-274

Cleary L (1977) Succès de reproduction du cormoran à aigrettes, Phalacrocorax auritus auritus, sur trois Îles du St Laurent, en 1975 et 1976. MSc Thesis, L'Université Laval, pp 1-68

Collins MS, Bashiruddin JB, Alexander DJ (1993) Deduced amino acid sequences at the fusion protein cleavage site of Newcastle disease viruses showing variation in antigenicity and pathogenicity. Arch Virol 128(3-4):363-370

Czeglédi A, Ujvàri D, Somogyi E et al (2006) Third genome size category of avian paramyxovirus serotype 1 (Newcastle disease virus) and evolutionary implications. Virus Res 120(1-2):36-48

Dawson, PS (1973) Epidemiological aspects of Newcastle disease. Bull OIE 79, 27-34

Doyle TM (1927) A hitherto unrecorded disease of fowls due to a filter-passing virus. J Comp Pathol Therapeut 40:144-169

Doyle TM (1935) Newcastle disease of fowls. J Comp Pathol Therapeut 48:1-20

Erickson GA, Maré CJ, Gustafson GA et al (1977) Interactions between viscerotropic velogenic Newcastle disease virus and pet birds of six species. I. Clinical and serologic responses, and viral excretion. Avian Dis 21(4):642-654

Guittet M, Le Coq H, Morin M et al (1993) Proceedings of the Tenth World Veterinary Poultry Association Congress, Sydney, p 179

Halasz F (1912) Contributions to the knowledge of fowlpest. Veterinary Doctoral Dissertation, Communications of the Hungarian Royal Veterinary School, Patria, Budapest pp 1-36

Hannoun C (1977) Isolation from birds of influenza viruses with human neuraminidase. Dev Biol Stand 39:469-472

Heckert RA (1993) Ontario. Newcastle disease in cormorants. Can Vet J 34(3):184

Herczeg J, Pascucci S, Massi P et al (2001) A longitudinal study of velogenic Newcastle disease virus genotypes isolated in Italy between 1960 and 2000. Avian Pathol 30:163-168

Herczeg J, Wehmann E, Bragg RR et al (1999) Two novel genetic groups (VIIb and VIII) responsible for recent Newcastle disease outbreaks in Southern Africa, one (VIIb) of which reached Southern Europe. Arch Virol, 144(11):2087–2099

Huang Z, Panda A, Elankumaran S et al (2004) The hemagglutinin-neuraminidase protein of Newcastle disease virus determines tropism and virulence. J Virol 78(8):4176-4184

Hugh-Jones M, Allan WH, Dark FA, Harper GJ (1973) The evidence for the airborne spread of Newcastle disease. J Hyg 71(2):325-339

Jørgensen PH, Handberg KJ, Ahrens P et al (1999) An outbreak of Newcastle disease in free-living pheasants (Phasianus colchicus). Zentralbl Veterinarmed B 46(6):381-387

Kaleta EF, Baldauf C (1988) Newcastle disease in free-living and pet birds. In: Alexander DJ (ed) Newcastle disease. Kluwer Academic, Boston, pp 197-246

Kinde H, Uzal F, Hietala S et al (2003) The diagnosis of exotic Newcastle disease in southern California: 2002-2003. Proceedings of the 46th Annual Conference of the American Association of Veterinary Laboratory Diagnosticians San Diego, CA, October 11-13 2003

Kraneveld FC (1926) A poultry disease in the Dutch East Indies. Nederlands-Indische Bladen voor Diergeneeskunde 38:448-450

Kuiken T (1998) Newcastle disease and other causes of mortality in double-crested cormorants (Phalacrocorax auritus). PhD Thesis University of Saskatchewan, 174 p

Lamb RA, Collins PL, Kolakofsky D et al (2005) Family Paramyxoviridae. In: Fauquet CM, Mayo MA, Maniloff J et al (eds) Virus taxonomy, Eighth Report of the International Committee on Taxonomy of Viruses. Elsevier, San Diego, pp 655-668

Lancaster JE (1966) Newcastle disease - a review 1926-1964. Monograph no 3, Canada Department of Agriculture, Ottawa

Lancaster JE, Alexander DJ (1975) Newcastle disease: virus and spread. Monograph no 11, Canada Department of Agriculture, Ottawa

de Leeuw OS, Hartog L, Koch G, Peeters BP (2003) Effect of fusion protein cleavage site mutations on virulence of Newcastle disease virus: non-virulent cleavage site mutants revert to virulence after one passage in chicken brain. J Gen Virol 84(Pt 2):475-484

Lomniczi B, Wehmann E, Herczeg J et al (1998) Newcastle disease outbreaks in recent years in Western Europe were caused by an old (VI) and a novel genotype (VII). Arch Virol 143(1):49-64

Macpherson LW (1956) Some observations on the epizootiology of Newcastle disease. Can J Comp Med 20(5):155-168

McFerran JB, McCracken RM (1988) Newcastle disease. In: Alexander DJ (ed) Newcastle Disease, Kluwer Academic, Boston, pp 161-183

Mebatsion T, Verstegen S, De Vaan LT et al (2001) A recombinant Newcastle disease virus with low-level V protein expression is immunogenic and lacks pathogenicity for chicken embryos. J Virol 75(1):420-428

Mixson MA, Pearson JE (1992) Velogenic neurotropic Newcastle disease (VNND) in cormorants and commercial turkeys FY 1992. In: Proceedings of the 96th Annual Meeting of the United States Animal Health Association, Louisville, Kentucky, 1992, pp 357-360

OIE (2007) List of countries by disease situation http://wwwoieint/wahid-prod/publicphp?page=disease_status_lists accessed 20th September 2007

Panigrahy B, Senne DA, Pearson JE et al (1993) Occurrence of velogenic viscerotropic Newcastle disease in pet and exotic birds in 1991. Avian Dis 37(1):254-258

Parede L, Young PL (1990) The pathogenesis of velogenic Newcastle disease virus infection of chickens of different ages and different levels of immunity. Avian Dis 34(4):803-808

Pedersden KA, Sadasiv EC, Chang PW, Yates VJ (1990) Detection of antibody to avian viruses in human populations. Epidemiol Infect 104:519-525

Peeters BP, de Leeuw OS, Koch G, Gielkens AL (1999) Rescue of Newcastle disease virus from cloned cDNA: evidence that cleavability of the fusion protein is a major determinant for virulence. J Virol 73(6):5001-5009

Pospisil Z, Zendulkova D, Smid B (1991) Unexpected emergence of Newcastle disease virus in very young chicks. Acta Vet Brno 60:263-270

Römer-Oberdörfer A, Veits J, Werner O, Mettenleiter TC (2006) Enhancement of pathogenicity of Newcastle disease virus by alteration of specific amino acid residues in the surface glycoproteins F and HN. Avian Dis 50(2):259-263

Rott R, Klenk HD (1988) Molecular basis of infectivity and pathogenicity of Newcastle disease virus. In: Alexander, DJ (ed) Newcastle disease, Kluwer Academic, Boston, pp 98-112

Senne DA, Pearson JE, Miller LD, Gustafson GA (1983) Virus isolations from pet birds submitted for importation into the United States. Avian Dis 27(3):731-744

Shengqing Y, Kishida N, Ito H et al (2002) Generation of velogenic Newcastle disease viruses from a nonpathogenic waterfowl isolate by passaging in chickens. Virology 301(2):206–211

Utterback WW, Schwartz JH (1973) Epizootiology of velogenic viscerotropic Newcastle disease in southern California, 1971-1973. J Am Vet Med Assoc 163(9):1080-1088

Westbury H (2001) Commentary Newcastle disease virus: an evolving pathogen. Avian Pathol 30:5-11

Wobeser G, Leighton FA, Norman R et al (1993) Newcastle disease in wild water birds in western Canada. Can Vet J 34(6):353-359

3 Notification of Avian Influenza and Newcastle Disease to the World Organisation for Animal Health (OIE)

Antonio Petrini and Bernard Vallat

3.1 The World Organisation for Animal Health (OIE) and Its Mission

The World Organisation for Animal Health (OIE) is an intergovernmental organisation created by an international agreement in 1924. In May 2007, the OIE consisted of 169 member countries. The OIE develops standards for the use of these countries to protect themselves from disease incursion while avoiding unjustified sanitary barriers. These standards are scientifically based and are prepared by elected specialist commissions and working groups comprising world renowned scientific experts in the relevant fields. Most of these experts belong to the OIE worldwide network of 190 collaborating centres and reference laboratories. The standards are adopted by the General Assembly of Member Countries, which meets annually in May in Paris. OIE standards are recognised as sanitary international references by the World Trade Organisation (WTO). One of the main missions of the OIE is to ensure transparency in the global animal-disease situation. Each member country undertakes to report animal diseases detected on its territory. The OIE then disseminates the information to other countries, which can take the necessary preventive action. This information also includes diseases transmissible to humans and intentional introduction of pathogens. Information is sent out immediately or periodically depending on the seriousness of the disease. This objective applies to diseases that are naturally occurring as well as those that are deliberately caused. In order to fulfil its mandate in this respect, the OIE manages the World Animal Health Information System (WAHIS). Access to this application is restricted and allows users from member countries, namely delegates or their nominees, to electronically submit standard notification reports (immediate notification and follow-up reports, 6-monthly reports and annual reports) to the OIE. This system not only provides countries with a simpler and quicker method of sending notifications and reports on disease information but also allows them to benefit from the new analysis capabilities put in place to produce essential and useful information without delays. The World Animal Health Information Database (WAHID) interface provides public access to all data held within the WAHIS.

3.2 OIE Single Disease List

In 2001, the OIE was instructed to establish a single list of animal diseases that replaced lists A and B. In 2004, science-based criteria for listing a disease in the disease list (OIE 2007c) were defined and approved by member countries (Table 3.1). The aim was to develop criteria that would be acceptable to all member countries and to ensure that, if a disease was considered to be very important, it would be listed. The overriding criterion for a disease to be listed is its potential for international spread. Thereafter, other criteria are also examined, such as zoonotic potential or capacity for significant spread within naïve populations. Within each criterion, there are measurable parameters. If a disease fulfils at least one of these parameters and meets one or more criteria, in addition to its potential for international spread, then the disease is added to the OIE list.

Emerging diseases may be included in the OIE list if they have a zoonotic impact and/or a significant impact on mortality and/or morbidity within a naïve population. Dealing with emerging diseases this way avoids their international spread to other areas and regions. This possibility of introducing new emerging diseases will assist in addressing these diseases more efficiently in the future–with the knowledge that

I. Capua, D.J. Alexander (eds.) *Avian Influenza and Newcastle Disease*,

Table 3.1 Criteria for listing a disease in the OIE disease list

Basic criteria (always considering 'worst case' scenario)	Parameters (at least one 'yes' answer means that the criterion has been met)
International spread	Has international spread been proven on three or more occasions? OR Are more than three countries with populations of susceptible animals free of the disease or facing impending freedom (based on Code provisions,especially Appendix 3.8.1)? OR Do OIE annual reports indicate that a significant number of countries with susceptible populations have reported absence of the disease for several consecutive years?
Zoonotic potential	Has transmission to humans been proven? (with the exception of artificial circumstances) AND Is human infection associated with severe consequences? (death or prolonged illness)
Significant spread within naïve populations	Does the disease exhibit significant mortality at the level of a country or compartment? AND/OR Does the disease exhibit significant morbidity at the level of a country or compartment?
Emerging diseases (a newly recognised pathogen or known pathogen behaving differently)	Is there rapid spread with morbidity/mortality and/or apparent zoonotic properties?

their number will certainly increase as a consequence of globalisation, urbanisation, climate change, etc.

Highly pathogenic avian influenza (AI) in birds and low pathogenicity notifiable avian influenza in poultry (LPNAI) (as defined in Chapter 2.7.12 of the *Terrestrial Animal Health Code*) as well as Newcastle disease (ND) are included in the OIE list within the category of avian diseases.

3.3 Notification and Epidemiological Information

3.3.1 Immediate Notification and Follow-up

Under the OIE notification system, not only diseases but also infections without clinical signs and other significant epidemiological events require urgent notification within 24 h by a member country. The events of epidemiological significance must be reported immediately to the OIE, as stipulated in Article 1.1.2.3 of the chapter entitled 'Notification and Epidemiological Information' in the *Terrestrial Animal Health Codes* (OIE 2007b).

Events of epidemiological significance that must be reported immediately by member countries are as follows:

- The first occurrence of an OIE-listed disease or infection in a country or zone/compartment
- The re-occurrence of a listed disease or infection in a country or zone/compartment following a report by the delegate of the member country declaring the previous outbreak(s) eradicated
- The first occurrence of a new strain of a pathogen of a listed disease in a country or zone/compartment
- A sudden and unexpected increase in morbidity or mortality caused by an existing listed disease
- An emerging disease with significant morbidity/mortality or zoonotic potential
- Evidence of a change in the epidemiology of a listed disease (e.g. host range, pathogenicity, strain of causative pathogen), particularly if the disease has a zoonotic impact.

Most requirements for new criteria were aimed at diseases formerly included in list A. By establishing a single list, notification requirements have been extended to a longer list of diseases. In addition, the new obligations of member countries clearly address the concept of infections in which clinical manifestations of the disease are absent. The new obligations of member countries are better adapted to emerging diseases, including those of a zoonotic nature, and go even further by addressing a new concept, i.e. to report an emerging event even if the aetiological agent is unknown or has not yet been identified.

3.4 Avian Influenza

The definition of "notifiable avian influenza" (NAI), as described in Chapter 2.7.12 of the *Terrestrial Animal Health Code* (OIE 2007d) and adopted by the OIE International Committee during its 75th General Session is as follows:

"For the purposes of international trade, avian influenza in its notifiable form (NAI) is defined as an infection of poultry caused by any influenza A virus of the H5 or H7 subtypes or by any AI virus with an intravenous pathogenicity index (IVPI) greater than 1.2 (or as an alternative at least 75% mortality) as described below. NAI viruses can be divided into highly pathogenic notifiable avian influenza (HPNAI) and LPNAI:

a) HPNAI viruses have an IVPI in 6-week-old chickens greater than 1.2 or, as an alternative, cause at least 75% mortality in 4- to 8-week-old chickens infected intravenously. H5 and H7 viruses which do not have an IVPI of greater than 1.2 or cause less than 75% mortality in an intravenous lethality test should be sequenced to determine whether multiple basic amino acids are present at the cleavage site of the haemagglutinin molecule (HA0); if the amino acid motif is similar to that observed for other HPNAI isolates, the isolate being tested should be considered as HPNAI;
b) LPNAI are all influenza A viruses of H5 and H7 subtype that are not HPNAI viruses".

In the same chapter "poultry" is defined as:

"...all domesticated birds, including backyard poultry, used for the production of meat or eggs for consumption, for the production of other commercial products, for restocking supplies of game, or for breeding these categories of birds, as well as fighting cocks used for any purpose'. Birds that are kept in captivity for any reason other than those reasons referred to in the preceding paragraph, including those that are kept for shows, races, exhibitions, competitions, breeding or selling these categories of birds as well as pet birds, are not considered to be poultry".

The chapter deals not only with the occurrence of clinical signs caused by NAI virus, but also with the presence of infection with NAI virus in the absence of clinical signs.

The presence of AI viruses in wild birds creates a particular problem. In essence, no country can declare itself free from AI carried in wild birds. However, the definition of NAI refers to infection in poultry only and that 'for the purposes of international trade, a country should not impose immediate trade bans in response to a notification of infection with HPAI and LPAI virus in birds other than poultry'.

Regarding the presence of antibodies: 'Antibodies to H5 or H7 subtype of NAI virus which have been detected in poultry and are not a consequence of vaccination have to be further investigated.' In the case of isolated serological positive results, NAI infection may be ruled out on the basis of a thorough epidemiological investigation that does not demonstrate further evidence of NAI infection.

The following defines the occurrence of infection with NAI virus:

a) HPNAI virus has been isolated and identified as such or viral RNA specific for HPNAI has been detected in poultry or a product derived from poultry; or
b) LPNAI virus has been isolated and identified as such or viral RNA specific for LPNAI has been detected in poultry or a product derived from poultry.

Standards for diagnostic tests, including pathogenicity testing, are described in the *Terrestrial Manual* (OIE 2004b). Any vaccine used should comply with the standards described in the manual (OIE 2004c).

The NAI status of a country, zone or compartment can be determined on the basis of the following criteria:

- The outcome of a risk assessment identifying all potential factors for NAI occurrence and their historic perspective.
- NAI is notifiable in the whole country, an on-going NAI awareness programme is in place and all notified suspect occurrences of NAI are subjected to field and, where applicable, laboratory investigations.
- Appropriate surveillance is in place to demonstrate the presence of infection in the absence of clinical signs in poultry and to determine the risk posed by birds other than poultry; this may be achieved through an NAI surveillance programme in accordance with Appendix 3.8.9 of the *Terrestrial Code* (OIE 2007a).

3.5 Newcastle Disease

It seems likely that the vast majority of birds are susceptible to infection with ND viruses of both high and low virulence for chickens, although the clinical signs seen in birds infected with ND virus vary widely and are dependent on factors such as the virus, host species, age of host, infection with other organisms, environmental stress and immune status. In some circumstances, infection with extremely virulent viruses may result in sudden high mortality despite comparatively few clinical signs.

Even for susceptible hosts, such as chickens, ND viruses show a considerable range of virulence. Generally, variation consists of clusters around the two extremes, as shown in tests used to assess virulence; but, for a variety of reasons, some viruses may show intermediate virulence.

The enormous variation in virulence and clinical signs means that it is necessary to carefully define what constitutes ND for the purposes of trade, control measures and policies. The OIE definition (OIE 2004c) for reporting an outbreak of ND is reported in Chapter 2.

Newcastle disease virus has a high risk of spread from the laboratory; consequently, a risk assessment should be carried out to determine the level of biosecurity needed for the diagnosis and characterisation of the virus. Countries lacking access to such a specialised national or regional laboratory should send specimens to an OIE Reference Laboratory.

According to the chapter on Newcastle Disease of the *Terrestrial Code* (OIE 2007e), a country may be considered free from ND when it has been shown that ND has not been present for at least the past 3 years. This period shall be 6 months after the slaughter of the last affected animal for countries in which a stamping-out policy is practised, with or without vaccination against ND.

The Veterinary Administrations of ND-free countries may prohibit importation or transit of the following commodities through their territory from countries considered infected with ND:

- Domestic and wild birds
- Day-old birds
- Hatching eggs
- Semen of domestic and wild birds
- Fresh meat of domestic and wild birds
- Meat products of domestic and wild birds that have not been processed to ensure destruction of the ND virus
- Products of animal origin (from birds) intended for use in animal feeding or for agricultural or industrial use.

References

Office International des Épizooties (OIE) (2004a) Chapter 1.1.7 Principles of veterinary vaccine production. In: Manual of Diagnostic Tests and Vaccines for Terrestrial Animals. http://www.oie.int/eng/normes/manual/A_00018.htm

Office International des Épizooties (OIE) (2004b) Chapter 2.7.12 Avian Influenza. In: Manual of Diagnostic Tests and Vaccines for Terrestrial Animals http://www.oie.int/eng/normes/mmanual/A_00037.htm

Office International des Épizooties (OIE) (2004c) Chapter 2.7.15 Newcastle Disease. In: Manual of Diagnostic Tests and Vaccines for Terrestrial Animals. http://www.oie.int/eng/normes/mmanual/A_00038.htm

Office International des Épizooties (OIE) (2007a) Appendix 3.8.9 Guidelines for the surveillance of avian influenza. In: Terrestrial animal health code, 16th edn. OIE, Paris

Office International des Épizooties (OIE) (2007b) Chapter 1.1.2 Notification and epidemiological information. In: Terrestrial animal health code, 16th edn. OIE, Paris

Office International des Épizooties (OIE) (2007c) Chapter 2.1.1 Criteria for listing diseases. In: Terrestrial animal health code, 16th edn. OIE, Paris

Office International des Épizooties (OIE) (2007d) Chapter 2.7.12 Avian Influenza. In: Terrestrial animal health code, 16th edn. OIE, Paris Office International des Épizooties (OIE) (2007e) Chapter 2.7.13 Newcastle disease. In: Terrestrial animal health code, 16th edn. OIE, Paris

4 Emergency Response on Suspicion of an Avian Influenza or Newcastle Disease Outbreak

Manuela Dalla Pozza and Stefano Marangon

4.1 Introduction

Emergency management of animal diseases is required when there are outbreaks of contagious diseases in a country or region that is free of infection. Under these circumstances, animal diseases have the potential to cause serious socio-economic consequences in a given country or area. Such diseases may also spread easily and reach epidemic proportions, thus requiring joint and coordinated interventions between several countries.

The prompt identification of an infectious disease is a prerequisite for the appropriate management of such emergency situations. This is, however, useful only if the infrastructure can respond quickly and adequately to the emergency situation, so that all the necessary measures to contain and then progressively eliminate the infection are implemented.

Thus, rapid eradication of highly contagious animal diseases depends on the preparedness to respond to its introduction into a disease-free area and on the way this response is implemented from the time of suspicion. Avian influenza (AI) and Newcastle disease (ND) are among the major concerns of animal husbandry organisations, due to the economic losses these diseases cause to the poultry industry as a result of illness and death of susceptible animals, disruptions in the marketing system following the implementation of restriction policies and, in the case of AI, the potential human health implications.

The management of a suspected index case of highly pathogenic AI or ND is crucial to subsequent actions aimed at limiting the spread of infection and, ultimately, in prompt eradication of the virus. It includes actions at local veterinary headquarters and at the farm level, which must be coordinated centrally to ensure the flow of information that is essential for educated decision-making. The success of any emergency intervention strategy is dependent on the level of preparedness, including action plans to source manpower and equipment, as well as on the degree of communication between relevant parties. The availability of information on the successes and failures in field outbreak management will inevitably result in improved intervention strategies worldwide.

4.2 Preparedness for AI Outbreak Management

In the event of an outbreak, a disease control strategy should be adopted, in line with the international obligations. Every country's objective in tackling outbreaks of any disease is to restore a disease-free status as quickly as possible. A contingency plan should be available, specifying the national measures to be implemented in the event of an outbreak. The contingency plan represents an important instrument in the management of a disease emergency and should include two main documents: a resource plan and an up-to-date operational manual. The latter should describe, in detail and in a comprehensive and practical manner, all the actions, procedures, instructions and control measures to be implemented in handling an outbreak of AI/ND. These guidelines should be made available at the farm level, to the farmer and the farm veterinarian (providing information on biosecurity and disease awareness), to competent veterinary authorities (providing information on staff, resources, equipment and facilities, instructions related to notification, slaughter and disposal of birds, compensation, disinfection, emergency vaccination and other related ac-

I. Capua, D.J. Alexander (eds.) *Avian Influenza and Newcastle Disease*,

tivities) and to diagnostic laboratories (providing information on required equipment, sample handling and diagnostic procedures). The operational manual ensures a straightforward road map to plan activities at different levels (farm, laboratory, crisis unit) from the time of notification of a clinical suspicion.

4.3 Action at the Time of Suspicion

As soon as the suspicion of AI or ND is reported to the appropriate authorities, the official veterinarian (OV) identifies the person who has reported the suspicion; if the latter is the farmer, the OV collects information concerning:

a) Location, characteristics and number of birds and other animals on the farm
b) Presence of staff and vehicles
c) Recent movement of people, equipment, vehicles and animals.

4.3.1 Reporting the Suspicion

It is also an obligation for the company veterinarian or private practitioner to support the OV in collecting information. If there is a delay between the suspicion and the arrival of the OV, the company veterinarian and/or private practitioner must do everything in his/her power to prevent the infection from spreading. The OV shall inform the local and regional veterinary officer, the local or regional epidemiology centre and the official laboratory of the suspicion of AI. Moreover, the OV identifies the closest mobile disinfection unit, indicating the suspicion of AI, and equips him/herself with a dedicated sampling kit.

4.3.2 Farm Visit

During the farm visit, all the precautions to avoid further spread of infection should be taken. The vehicles of the OV and company/field veterinarian must be left outside the infected premises and at a safe distance from the entrance of the farm. The OV must visit the farm using adequate equipment (protective clothing, sampling equipment, etc.; see list in Annex 1). The OV coordinates measures taken at the farm level, in order to avoid movement of people, animals, equipment and vehicles from the suspected premises. An initial appraisal of the situation on the farm also should be made. This first assessment in the suspected outbreak is aimed at rapidly identifying animal species and number of birds, their distribution on the farm, the history of disease and recent movements to and from the site of the suspected outbreak.

4.3.3 Disease Investigation

This initial appraisal should be followed by an accurate clinical examination of the affected birds in order to establish the clinical condition of the susceptible animals, including sick and suspect birds. The clinical investigation must be performed on all susceptible species on site, and it must begin from the clinically unaffected units. Particular attention must be paid to the vaccination history of the farm. All this information must be reported in the epidemiological inquiry.

All the birds present, *per species*, must be identified, and for each species identified an outbreak report must be drafted, containing:

- The number and location of birds in different pens and compartments of the farm
- The date of onset of clinical signs
- A description of clinical signs
- Reported percentage mortality.

Following the collection of preliminary data, a post-mortem examination of dead animals (or euthanised moribund animals) must be performed. Samples are then to be collected from organs exhibiting pathological lesions. Organs from different apparatuses should not be pooled, and if possible, cross-contamination due to excision of organs with single pair of scissors or pliers should be avoided. The samples collected are then to be packaged appropriately (in leakproof containers, wrapped in at least two plastic bags) to avoid dissemination of the infectious agent and then transported refrigerated to the laboratory. Culled animals may be transported in a sealed autoclavable plastic bag inserted inside a similar sealed bag. All samples must be carried to the laboratory inside a polystyrene box containing icepacks. The polystyrene box must be appropriately disinfected on the outside before it is removed from the premises. The samples must be accompanied by a form that contains the main information of the suspected outbreak.

4.3.4 Epidemiological Investigation

An accurate epidemiological enquiry has to be carried out, aiming at the collection of data and information to be used for the implementation of subsequent actions on the premise and on in-contact farms (see Annex 2). The epidemiological investigation has the ultimate objectives of establishing:

- Origin and date of infection
- Means by which infection could have been spread to other farms (movements of animals, vehicles and staff, infected farms located in close proximity to the suspected outbreak)
- Farms at risk of infection

Reliable and comprehensive epidemiological data can be collected more easily if a standardised questionnaire is available. This is to be prepared during disease-free periods and should include all the relevant information related to farm characteristics and risk factors for AI introduction and spread: type of farming and management, type and number of birds, presence of risk factors for AI, anamnesis and pathological and clinical findings.

Data on risk factors related to the introduction and spread of AI should be collected within the period at risk of virus introduction (PVI) and the period at risk of virus spread from the farm (PVS) and reported in the epidemiological investigation form. The critical date is determined as the earliest time the virus could have entered the infected premise and should be consistent with the maximum incubation period, defined by the OIE code. The PVI should be identified within 21 days prior to the onset of first clinical signs; the PVS theoretically can coincide with the PVI. In practice, the most dangerous period for viral spread occurs between one week prior to the onset of clinical signs and the date the farm was placed under restriction. The OV and farm veterinarian are requested to carefully fill in the epidemiological inquiry form, collecting information on risk factors present in the above-mentioned PVI and PVS. With reference to the epidemiological inquiry, it is essential that animal, people and vehicle movements resulting in access to the farm are reported and traced. The information obtained from the tracing will help to decide the extent of the area at risk and identify any additional suspected or contaminated premises. The following relevant data should be collected:

- Movements to and from infected and suspected premises for at least 21 days before the first report of unusual morbidity or mortality; obtaining this information is a foremost priority.
- Movements of birds, eggs, poultry products, feed, litter, waste, equipment, people and possibly other animal species.
- Movements of trucks (all vehicles, regardless of their contact with animals).
- People involved with feed delivery, vaccinating crews, catching crews, tradesmen, company service staff, veterinarians and relatives should be interviewed and a list compiled of all possible contacts for 3 days after visiting any premises under suspicion for AI and 7 for ND.

The epidemiological inquiry must be sent (possibly faxed) to the competent authorities as soon as it has been completed. The epidemiological team has the task to analyse data originating from the field outbreaks and should generate a road map for the OV that includes:

- Notation of the high-risk area in which intensive monitoring programmes should be implemented
- A list of contact holdings ranked according to the level of risk.

These units must be put under strict veterinary surveillance and visited at least on a daily basis. The reliability of the tracing exercise depends on the farmers' cooperation and on the effectiveness of the methods used to record the animals as well as movements on and off the outbreak premises.

4.3.5 Additional Action on the Suspected Premise

The OV must also review security arrangements and supervise the enforcement of the restriction measures. It is recommended to supply written instructions to the farmer, with a list of rules that are to be followed to avoid further spread of infection. Among these:

- Animals must not be moved from their pen
- Animals must not be moved on or off the farm
- Visitors are not allowed on site, except for essential services
- All visitors must be prepared for a complete change of clothing, boots, etc.
- Visitors must comply with disinfection protocols when leaving the premise.

The task force is also required to identify all means of access to the premise and these are to be put under control to limit the access of people and vehicles. In addition, the access routes should be equipped with disinfection points (disinfectant lagoons, mobile disinfection units, sprayers, etc.). Moreover, a preliminary assessment of the slaughter and disposal procedures that are applicable under those circumstances is useful during this phase (see Chapter 12).

4.3.6 Action at Headquarters

The suspected outbreak has to be notified to local and central authorities and the local contingency plan must be implemented. Whilst the results of laboratory investigations are awaited, it is necessary to plan for a crisis centre and to identify all necessary resources that are to be dedicated to management of the crisis. In addition, it is essential to identify the restriction zones ahead of time so that an educated decision-making process can support the control measures that are to be implemented. It is also essential to estimate the staff requirements for the implementation of control procedures, as the outcome of intervention strategies is dependent on the availability of skilled manpower resources, especially when eradication campaigns are pursued in densely populated livestock areas.

Culling and disposal methods (incineration, burial and treatment at rendering plants) must be reviewed in order to select the method which is most "fit for purpose", taking into account the following factors:

- Number of birds on site.
- Environmental impact of disposal.
- Area of land to be used for the disposal of carcases (hydrogeologic characteristics).
- Availability of skilled manpower and suitable equipment.
- Capacity of rendering plants and availability of suitable transport lorries.
- Costs and time needed to complete stamping-out procedures.

Once the method of disposal of carcases has been selected, the manpower and equipment required to complete stamping-out and sanitation procedures must be estimated and the staff (slaughtermen, contractors, etc.) alerted.

4.3.7 Exit from the Farm

Following the clinical visit and the collection of samples, the OV and farm veterinarian, in the designated changing room, disinfect their protective gear and collect all sterilisable equipment in an autoclavable bag, which is sealed and inserted into a second bag, which is disinfected externally. All single-use materials, sheets of paper, disposable gear and shoe covers are put inside a plastic bag that is left on site.

5 Necropsy Techniques and Collection of Samples

Calogero Terregino

5.1 Necropsy

5.1.1 Introduction

If samples are to be collected for virological examination, several pairs of scissors and forceps should be available, particularly if single-organ virus isolation is to be attempted. By using the same surgical instruments to collect organs from different apparatuses, cross-contamination of samples may occur. This also applies if birds from different species (or different premises) are to be processed during the same session. Operators must comply with biosafety guidelines as indicated in Annex 3.

5.1.2 Procedure

Dampening the plumage with a disinfectant solution is strongly recommended to limit the dispersion of infected dust and feathers. The bird must be placed on its back with the feet towards the operator (Fig. 5.1). Following dislocation of the coxofemoral joint, the skin over the abdomen should be cut and removed (Fig. 5.2). The superficial breast muscles should be examined to determine whether decreased muscle mass, paleness (anaemia), haemorrhages, congestion or bruising are present.

The abdominal muscles, ribs and coracoid bone may be cut with robust scissors (poultry shears) (Figs. 5.3) and removed from the chest to expose the internal organs and the chest cavity (Figs. 5.4, 5.5). The liver, lungs, heart and the air sacs can now be examined. The lungs must be gently removed from the ribcage (Fig. 5.6) with the trachea. A longitudinal section of the larynx, trachea and syrinx will facilitate accurate examination of the mucosa (Fig. 5.7) and the collection of samples by swabbing (Fig. 5.8).

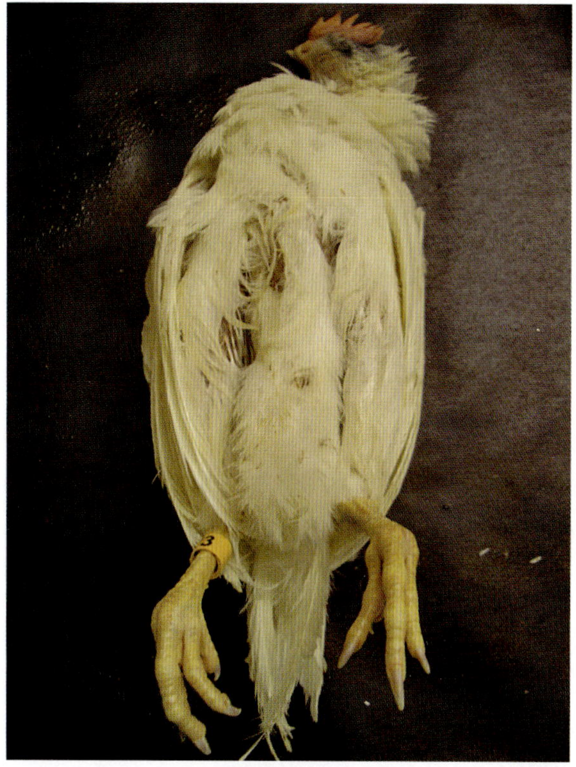

Fig. 5.1 Correct starting position for necropsy. The bird lies on its back with the feet towards the operator

The gastrointestinal canal may be excised between the oesophagus and proventriculus (Fig. 5.9a) and in the vicinity of the rectum near the cloaca (Fig. 5.9b). The proventriculus, gizzard and intestine can then be removed with the pancreas, liver and spleen. The spleen is a small, red, round organ located at the junction of the proventriculus and gizzard. The proventriculus and gizzard are to be cut open to detect the presence of feed (indicating that the bird has/has not suffered from anorexia) and for the presence of submucosal haemorrhages (Fig. 5.10). Examination of

I. Capua, D.J. Alexander (eds.) *Avian Influenza and Newcastle Disease*,

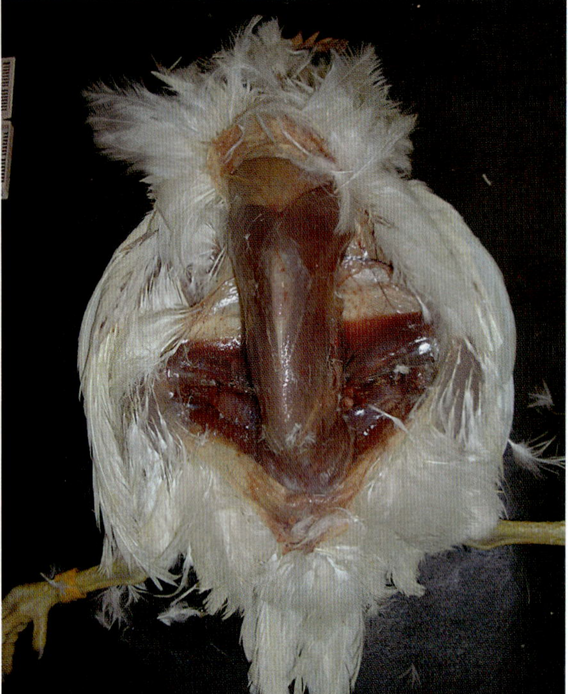

Fig. 5.2 The skin of the breast, abdomen and legs is reflected and the legs pulled and twisted to disarticulate the head of the femur from the acetabulum

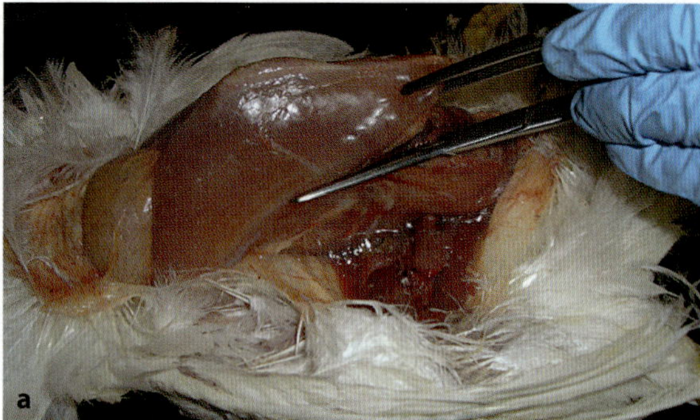

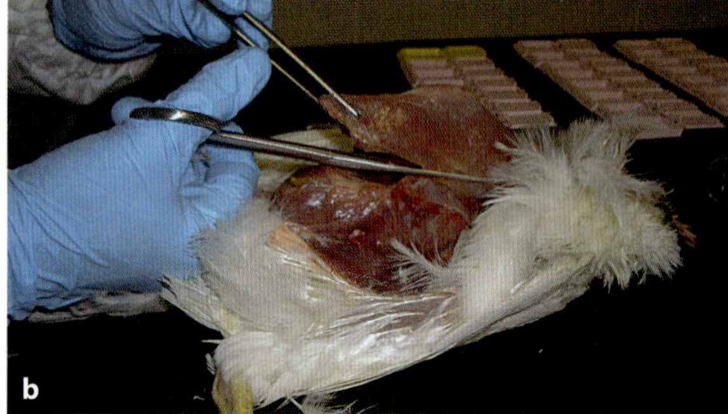

Fig. 5.3 a, b The sternum is incised along the ventral margin of the pectoral muscle continuing through the ribs. A similar incision is made on the opposite side of the breast (see panel b)

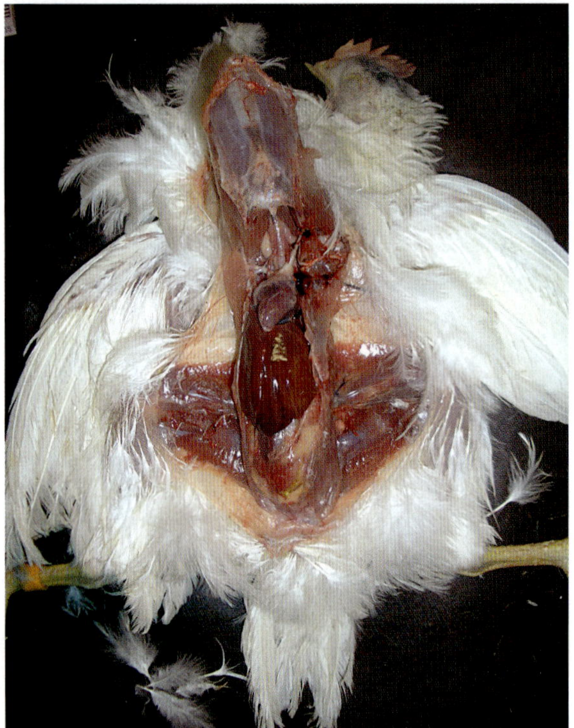

Fig. 5.4 View of internal organs after removal of the breast

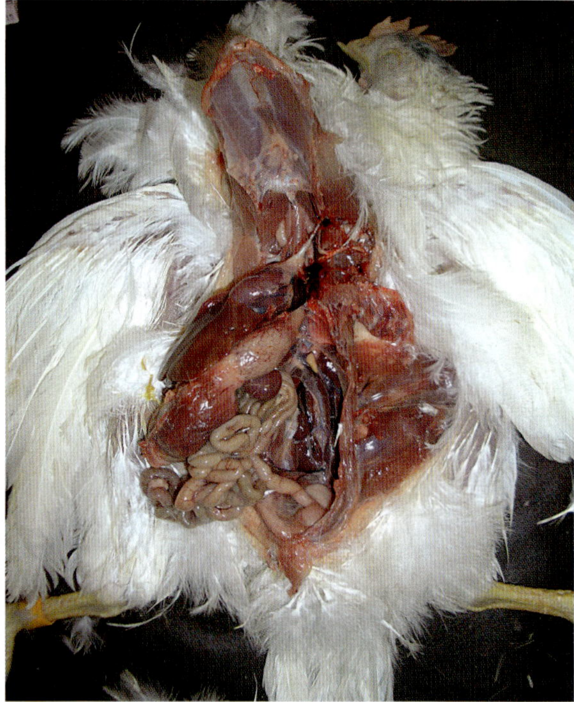

Fig. 5.5 View of body cavity after removal of the breast and lateral positioning of internal organs

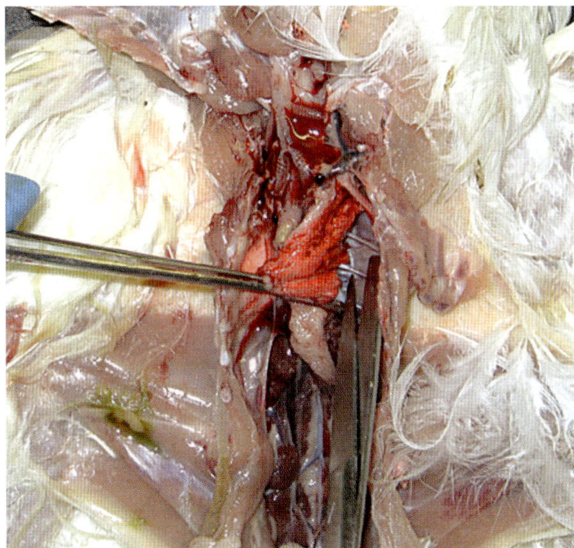

Fig. 5.6 Removal of the lung from the ribs

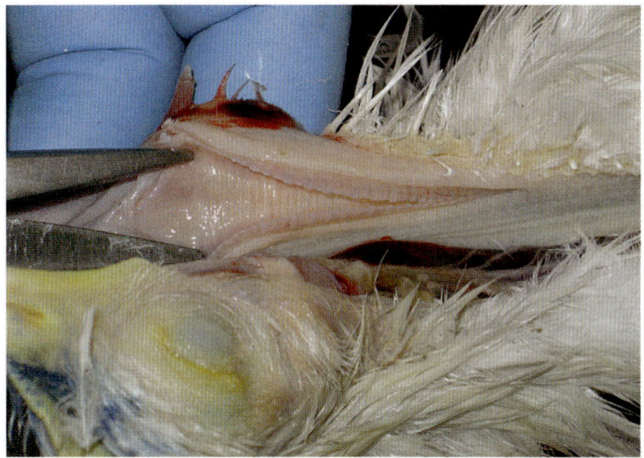

Fig. 5.7 Examination of tracheal mucosa

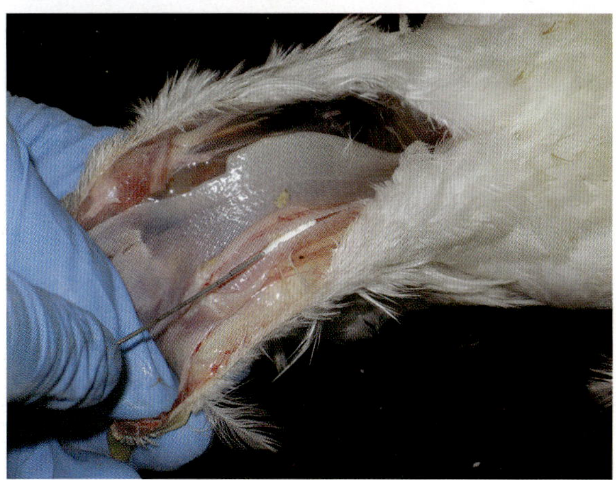

Fig. 5.8 Collection of a tracheal swab sample

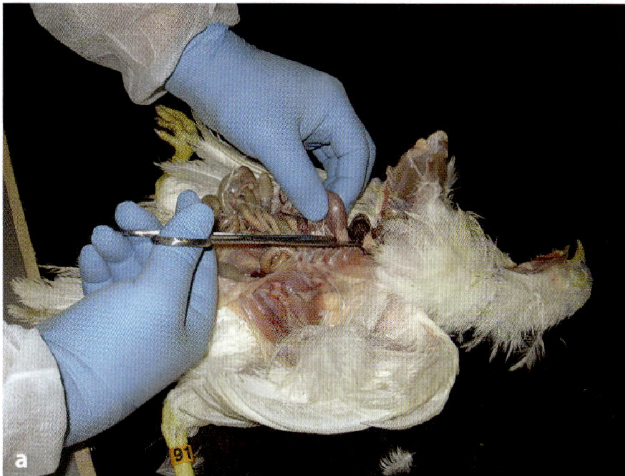

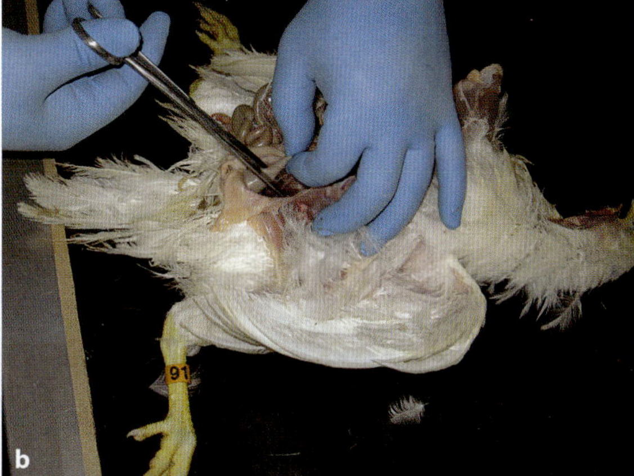

Fig. 5.9 a, b The intestinal viscera are removed by resection at the junction between oesophagus and proventriculus (see panel **a**) and by resection of the terminal region of the rectum near the cloaca (see panel **b**)

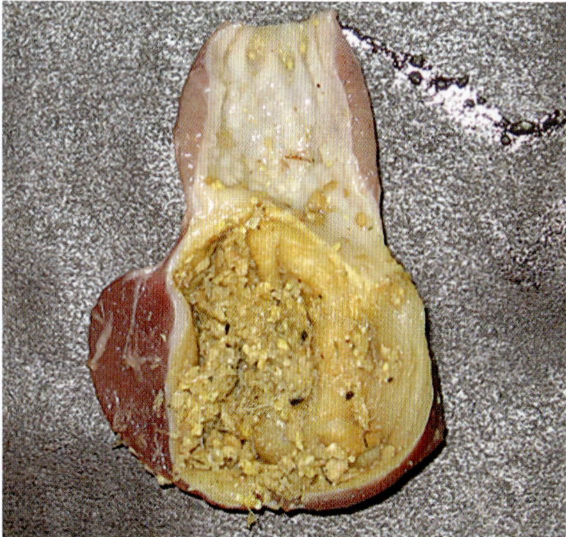

Fig. 5.10 Lumen of the proventriculus and gizzard

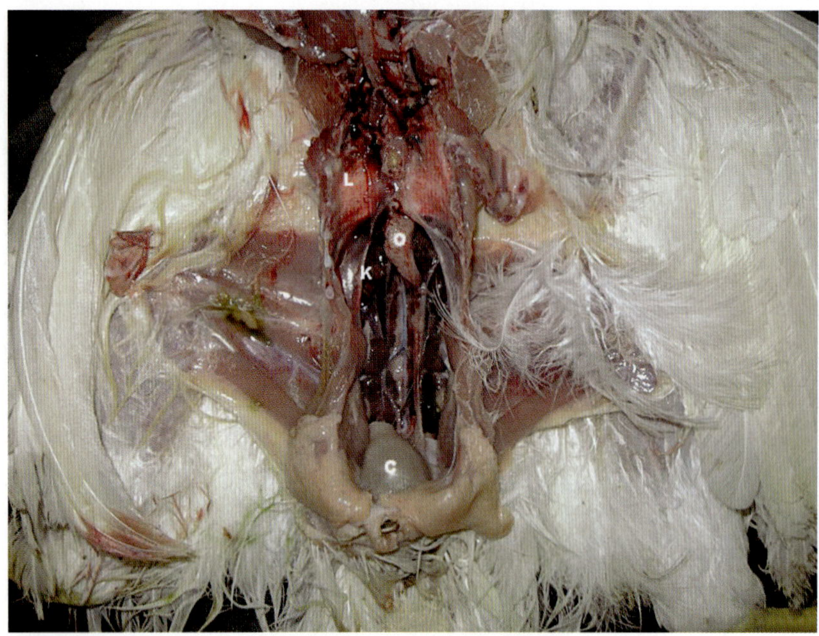

Fig. 5.11 View of internal organs after removal of the heart, digestive apparatus, liver and spleen. *L* lung, *O* ovary, *K* kidney and *C* cloaca

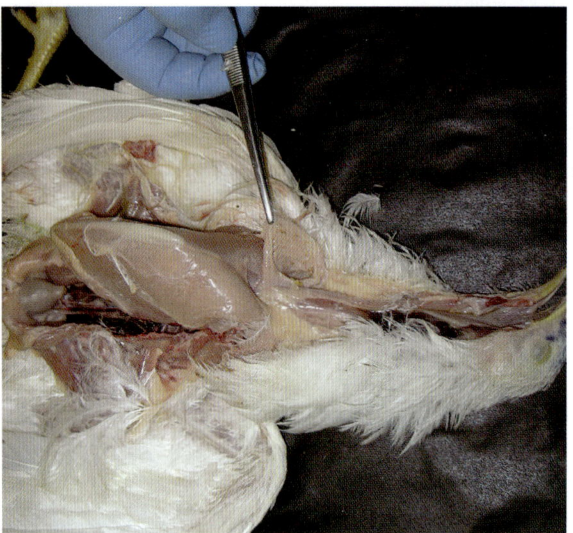

Fig. 5.12 Examination of the lumen of the crop

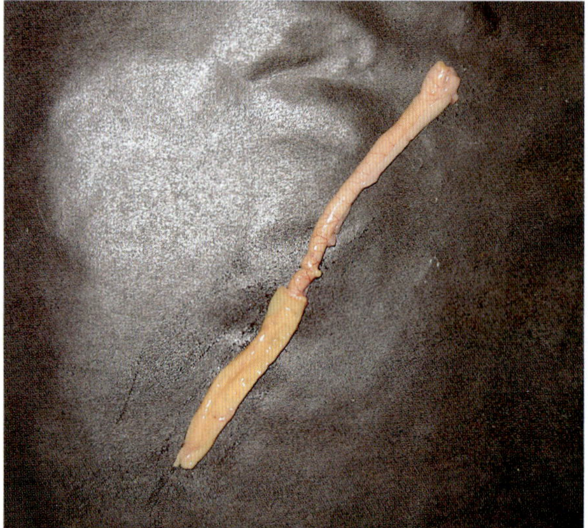

Fig. 5.13 Examination of duodenal lumen

the kidneys, (elongated and lobulated organs embedded in the vault of the pelvis) (Fig. 5.11) and observation of the ovaries and oviduct or paired testes, which are positioned on top of the kidneys, will follow. The necropsy is concluded with an examination and sampling of the mucosal surface of the oesophagus, crop (Fig. 5.12) and intestine (Fig. 5.13). Particular attention should be paid to lymphatic tissue, such as Peyer's patches and caecal tonsils, which may appear enlarged and haemorrhagic (Fig. 5.14).

To collect the brain, disarticulate and detach the head from the atlanto-occipital joint and raise the skin to expose the skull. Holding the skin with a hand, pull towards the beak (Fig. 5.15) and incise the skull from the occipital foramen forward in two

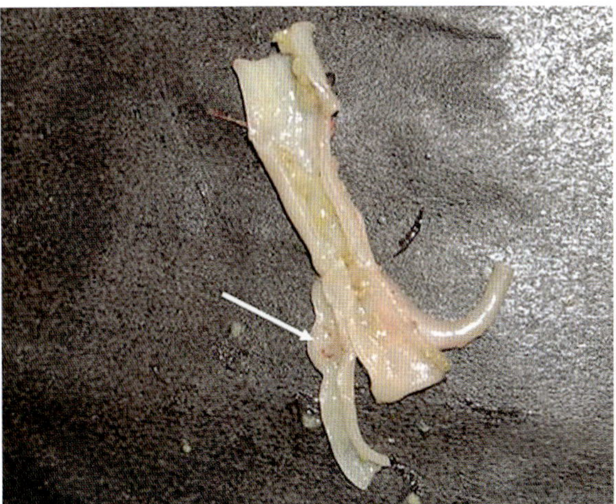

Fig. 5.14 Examination of the terminal part of the large intestine (rectum and caeca). A caecal tonsil is indicated by the *arrow*

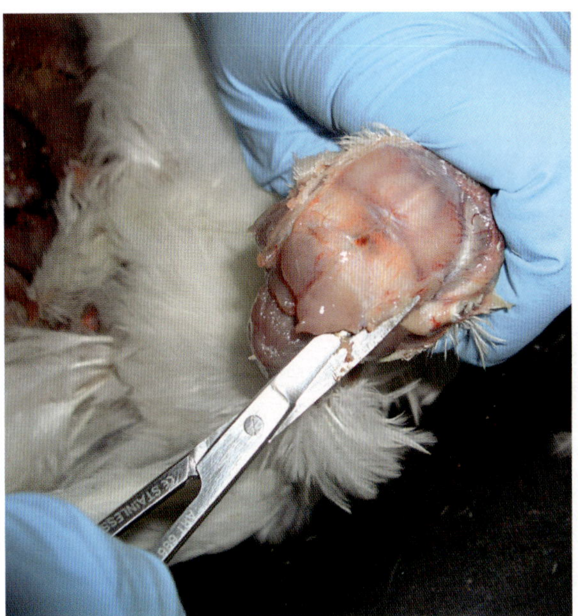

Fig. 5.15 Removal of skullcap by incision around the circumference of the skull

tangential cuts to both orbits, joining the incisions frontally. The cranial vault may be removed exposing both cerebral hemispheres, optic lobes and cerebellum (Fig. 5.16). To extract the brain, introduce delicately one arm of a pair of surgical forceps under the anterior region of the brain case and leverage gently upwards.

For a schematic view of chicken anatomy see Figure 5.17.

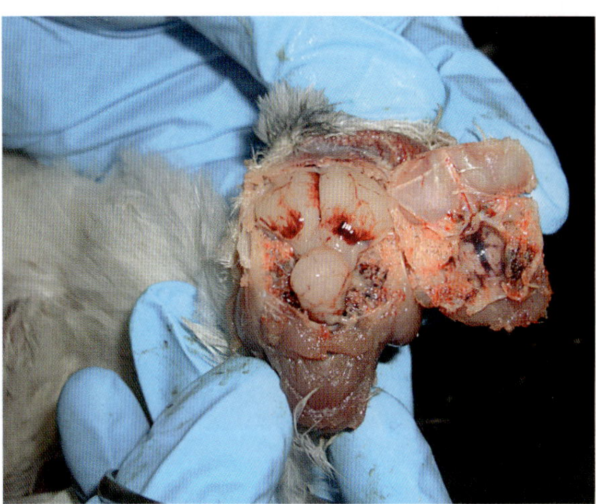

Fig. 5.16 View of brain after removal of the skullcap

5.2 Collection of Specimens from Live Animals and Carcases

It is crucial to select suitable specimens for laboratory investigation of suspected outbreaks of avian influenza or Newcastle disease. Decomposing carcases are of limited value for attempted virus isolation and detection. Ideally, sick as well as recently dead birds should be submitted to the laboratory for blood sampling, necropsy and diagnostic examination. Oropharyngeal/tracheal and cloacal swabs allow large numbers of samples to be taken on site from both live and dead birds for laboratory investigations.

In a suspected outbreak, 20 cloacal and 20 oropharyngeal swabs (or swabs from all birds if there are less than 20 in a flock) should be collected in each epidemiological unit. These should preferably be taken from birds showing signs of disease. It is important that the swabs are coated in faeces (optimum 1 g). If, for any reason it is impracticable to take cloacal swabs from live birds, fresh faeces samples may serve as an alternative.

A minimum of five birds should be collected for necropsy. Those birds exhibiting overt clinical signs should be selected for this purpose. 20 blood samples or samples from all birds if less than 20 in flock, should be collected in each epidemiological unit. Birds that are sick or apparently recovered should be targeted for blood sampling.

This sampling scheme has been developed to ensure a 99% probability of detecting at least one positive serum if 25% or more of the flock is positive, regardless of flock size.

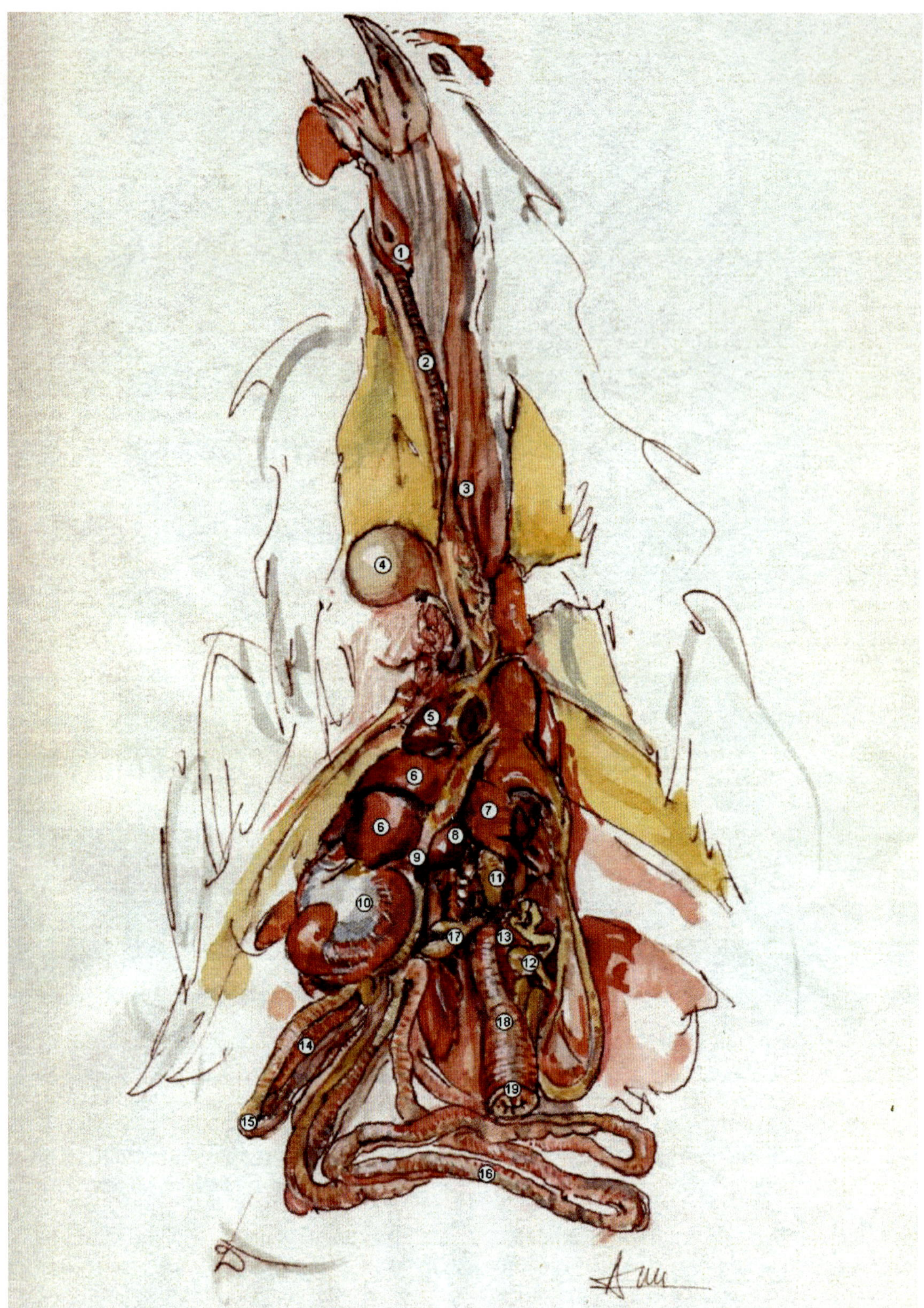

Fig. 5.17 Schematic view of chicken anatomy. *1* Larynx; *2* Trachea; *3* Oesophagus; *4* Crop; *5* Heart; *6* Liver; *7* Lung; *8* Spleen; *9* Proventriculus; *10* Gizzard; *11* Ovary; *12* Oviduct; *13* Kidney; *14* Pancreas; *15* Duodenum; *16* Small intestine; *17* Caecum; *18* Large intestine; *19* Cloaca. (Courtesy of Amelio Meini)

5.2.1 Blood

To obtain serum that will be tested for avian influenza or Newcastle disease antibodies, anticoagulant is not required and the blood is allowed to clot. In case of avian influenza, since highly pathogenic AI is an acute and lethal disease for most species, the detection of antibodies will only be helpful if low pathogenicity AI is suspected.

Materials

- *Test tubes*
- *2.5-ml syringes*
- *25-gauge needles for small birds (quail, partridge, parrots)*
- *23-gauge needles for larger birds (chicken, turkey)*
- *Cotton wool*
- *70% alcohol solution*

Collect the blood from the brachial vein (the largest vein under the wing) (Figs. 5.18, 5.19) as follows:

1. Place the bird on a table, setting it on its side.
2. Lift up the wing with one hand and part the feathers along the wing. Water or denaturated alcohol can be used to help keep the feathers separated.
3. Place the needle at a slight angle, bevel up, against the vein on the underside of the wing (the bevel is the side of the needle with the angle and the hole). Insert the needle into the vein and slowly draw blood.
4. Remove the needle and apply pressure to the vein for a few seconds. This will help to minimize the development of large haematomas, which occur commonly in poultry.
5. 1-2 mls of blood should be collected. Empty the syringe in an appropriate test tube with or without vacuum. In case of ordinary test tubes, the needle is to be removed from the syringe and the freshly drawn blood slowly collected through the tip of the syringe into the test tube. In case of vacuum-test tubes, insert the needle of the syringe containing the blood in the rubber stopper and allow blood to be drawn from vacuum. Avoid forcing blood through needle as this will causes haemolysis. Test tubes are to be positioned horizontally for 30-60 mins at room temperature to allow blood to clot. Serum should be separated from the blood clot and stored at +4°C or -20°C prior to testing.

Fig. 5.18 Chicken: brachial vein (wing vein). Feathers have been removed from the ventral surface

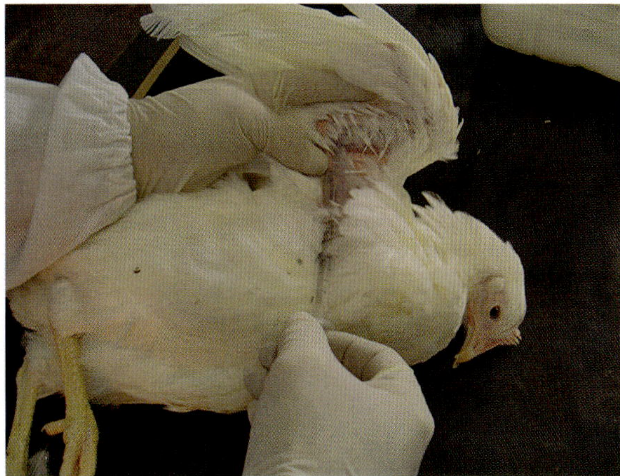

Fig. 5.19 Specific-pathogen free chicken: obtaining a blood sample from the brachial vein

In the absence of syringes, the following procedure can be used:

1. Pluck a few feathers from the bend in the wing to expose the median vein.
2. Using a sterile scalpel blade, prick the vein to obtain a blood sample.
3. Place the blood sample in a tube (2 ml is a sufficient amount).
4. To obtain serum, place the blood vial on a slanted surface for 10–15 min to allow for clotting. The serum samples can now be spun by centrifugation.
5. Allow the blood sample to stand for 4–12 h at room temperature.

Vials containing the blood samples should be refrigerated and sent to a diagnostic laboratory as soon as possible.

5.2.2 Tracheal/Oropharyngeal Swabs

Materials

- *Synthetic swabs (rayon and dacron) are preferred to cotton swabs, which may contain viral inhibitory or toxic substances*
- *Sterile tubes 5–15 ml*
- *Test-tube racks*
- *Virus transport media (VTM)*
- *PBS antibiotic solution.*

Protein-based media, such as brain-heart-infusion (BHI) or Tris-buffered tryptose broth, or other commercial viral transport media give added stability to the virus, especially during transport. The antibiotics used and their concentrations may be varied to suit local conditions and availability. Very high levels of antibiotics may be necessary for faecal samples. Recommended levels are: 10,000 IU penicillin/ml, 10 mg streptomycin/ml, 0.25 mg gentamycin/ml, and 5,000 IU nystatin/ml. These levels may be reduced by up to five-fold for tissues and tracheal swabs. BHI medium must be prepared in water and contain 15% w/v BHI broth powder, prior to sterilisation (by autoclaving at 121°C for 15 min). Following sterilisation, antibiotics must be added as follows: 10,000 IU penicillin/ml G, 20 µg amphotericin B and 1,000 µg gentamycin/ml. Media may be stored at 4°C for a maximum of 2 months.

Swab samples can be collected from dead or sick birds. Use dry swabs to collect samples from dead birds; swabs moistened with viral transport medium should be used for samples obtained from live birds. Insert the swab and rub the mucosa (Figs. 5.20 a-c). Use one swab for each bird. Place swabs into tubes containing enough VTM to moisten immerse and the swabs in 2 ml maximum of VTM (Fig. 5.21).

Swab samples collected from different birds may be pooled. Place a maximum of five swabs into one tube and immerse the swabs with 4–5 ml of VTM.

Record the necessary information (farm name, number of the sample, date) on the tube (not on the lid).

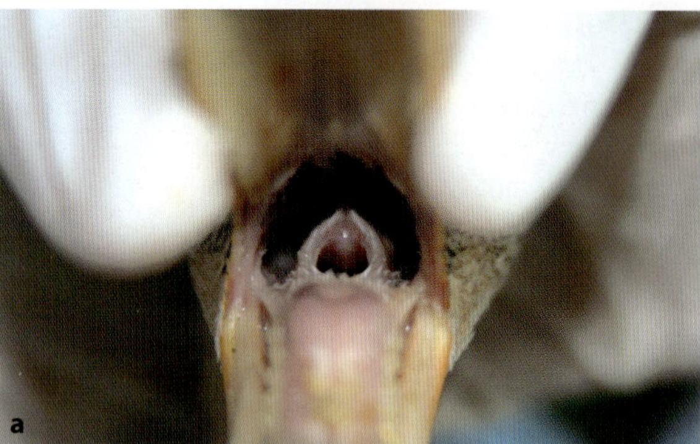

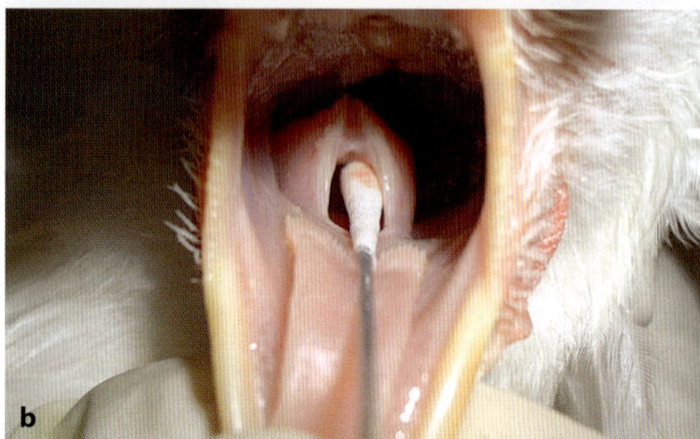

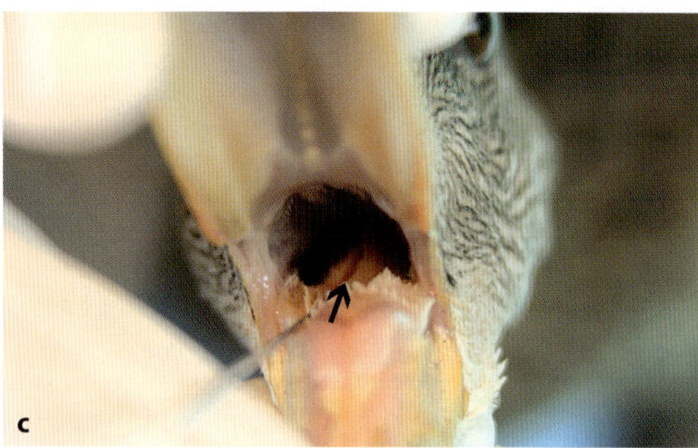

Fig. 5.20 a-c **a** Mallard (*Anas platyrhynchos*): larynx. **b** Specific-pathogen free chicken: introduction of a swab in the larynx. **c** Mallard (*Anas platyrhynchos*): swab in the trachea. The laryngeal opening is closed (*black arrow*)

5.2.3 Cloacal Swabs

Take swab samples from any dead or sick birds first, then swabs from the other birds. Use dry swabs to collect samples from dead birds; swabs moistened with VTM should be used for sample collections from live birds.

Insert the entire head of the swab into the cloaca (Figs. 5.22 a, b). Appling gentle pressure, rotate the swab inside the cloaca two or three times. Shake off large pieces of faeces and insert the swab into the tube. Immerse the swabs in VTM (2 ml maximum).

Swabs must be chilled immediately on ice or with frozen gel packs and submitted to the laboratory as

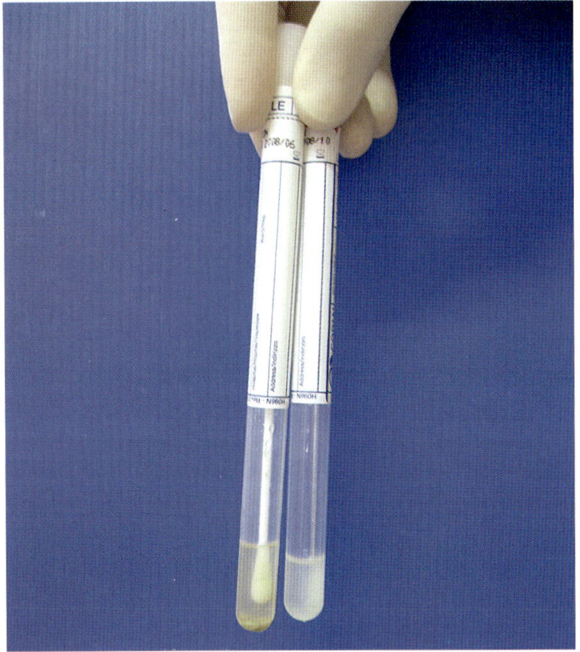

Fig. 5.21 Cloacal swab (*on the left*) and tracheal swab (*on the right*) in tubes containing viral transport media

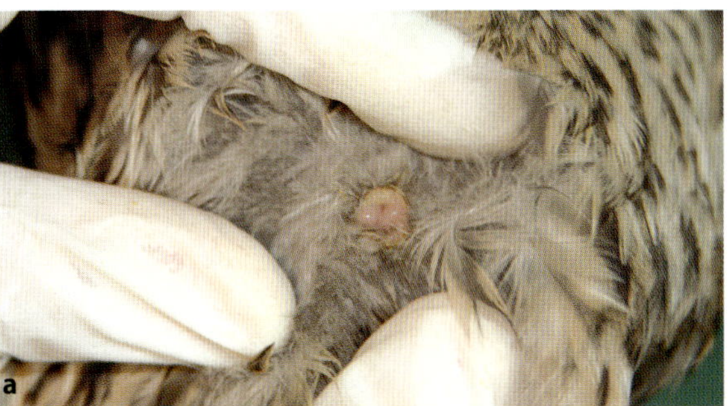

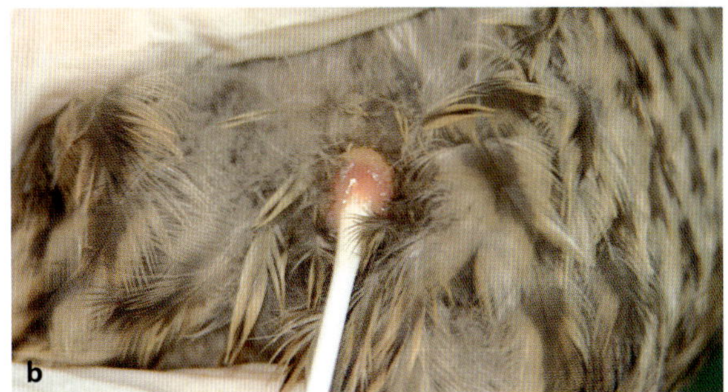

Fig. 5.22 a, b **a** Mallard (*Anas platyrhynchos*): cloacal orifice. **b** Mallard (*Anas platyrhynchos*): introduction of a swab in the cloacal orifice

quickly as possible. If rapid transport within 48 h to the laboratory is not guaranteed, samples must be immediately frozen, stored and then transported on dry ice. Swab samples collected from different birds may be pooled. Place a maximum of five swabs into one tube and immerse the swabs with 4-5 ml of VTM.

Swabs for PCR analysis can be sent dry provided they are dispatched rapidly to the lab and are transported under proper conditions.

5.2.4 *Birds for Necropsy*

A minimum of 5 whole carcases of birds that have died recently, or severely sick or moribund birds that have been killed humanely should be collected from each epidemiological unit. Put the carcases in doubled, resistant plastic bags and transport them in washing-resistant containers.

For attempted virus isolation, the following selection of organs can be collected in plastic rigid sterile containers:

- Respiratory tract (trachea, lungs)
- Intestine (duodena with pancreas, caecal tonsils)
- Liver (HPAI)
- Kidney (HPAI)
- Spleen (HPAI-ND)
- Brain (HPAI-ND)

All samples to be used for attempted virus isolation should be kept at approximately 4°C. If samples need to be despatched to a specialist diagnostic laboratory, it is essential that specimens are sent immediately on ice using an appropriate courier. If delays of > 2 days are expected, the samples should be frozen, ideally at –80°C, prior to being sent.

Clinical Traits and Pathology of Avian Influenza Infections, Guidelines for Farm Visit and Differential Diagnosis

6

Ilaria Capua and Calogero Terregino

6.1 Introduction

Influenza A viruses have been grouped into two distinct pathotypes on the basis of the severity of the disease they cause in susceptible chickens. Highly pathogenic avian influenza (HPAI) viruses cause severe disease in chickens and other gallinaceous birds. Low pathogenicity avian influenza (LPAI) viruses cause a much milder clinical disease. HPAI has been associated only with some strains of H5 and H7 (occasionally some H10) avian influenza viruses. The infection causes a lethal disease in most birds, particularly in galliforms (chickens and turkeys) and is characterised by very high flock mortality. Some birds, such as ostriches as well as domestic and wild waterfowl, may be resistant clinically to disease, although they harbour the virus and are permissive to viral replication. LPAI is caused by all 16 haemagglutinin subtypes (H1–H16) of the virus. Low pathogenicity strains are usually responsible for mild or inapparent disease. Viral replication is restricted to the respiratory, digestive and urogenital tracts; therefore, the clinical appearance of infection consists of respiratory signs, gastroenteric disorders and a decline in egg production. The severity of the clinical signs is influenced by several factors, such as age, species and health status of the bird, secondary infections, husbandry methods and the infecting viral strain. LPAI viruses are maintained in nature within the wild bird population, particularly in waterfowl and shorebirds.

6.2 Low Pathogenicity Avian Influenza in Poultry

6.2.1 Turkeys (*Meleagris gallopavo*)

6.2.1.1 Clinical Signs

Turkeys are extremely susceptible to AI and clinical signs reflect the extensive damage caused by the infection in the respiratory, digestive, urinary, and reproductive organs. Over the years, in both the Western and Eastern hemispheres, turkeys have been affected by a variety of subtypes, primarily H1, H5, H7, H6 and H9.

In meat turkeys, the severity of the clinical and post-mortem findings may vary significantly, with mortality ranging from a few percent to > 90%. This variability may be correlated with the age of the affected birds and with environmental factors such as hygienic condition of the farm, quality of ventilation and air, ambient temperature and presence of other infectious agents. The clinical condition and mortality appear to be more severe when the flock is also infected with *Mycoplasma* spp., or when bacteria. such as *Riemerella anatipestifer*, *Pasteurella multocida* or *Escherichia coli*, cause secondary infections. Similarly, secondary or contemporary infections with viral pathogens, including haemorrhagic enteritis virus (HEV), avian paramyxovirus 2 (APMV2), Newcastle disease (ND) vaccine viruses, adenoviruses, avian pneumoviruses and reoviruses, may result in a higher mortality.

Both the severity of clinical signs and the recovering capacity are age-related. In birds over 40 days of age, severe clinical signs generally regress, with recovery of most of the affected birds within a week from the onset of illness. In birds up to 40 days of age, the clinical condition often evolves into a more severe respiratory syndrome and may be associated with a mortality of > 20%. In certain cases, the surviving birds may remain stunted and do not reach optimal weights.

The appearance of clinical signs is generally associated with depression, ruffled feathers (Fig. 6.1), reluctance to move and absence of vocalisations inside the shed. The birds are still and the younger birds crowd under the heat lamps. Feed consumption drops

I. Capua, D.J. Alexander (eds.) *Avian Influenza and Newcastle Disease*,

Fig. 6.1 28-day-old turkeys naturally infected with low pathogenicity avian influenza (LPAI) of the H7N1 subtype during the acute phase of the disease. Depression, ruffled feathers and conjunctivitis are seen in the affected birds

Fig. 6.2 28-day-old poult naturally infected with LPAI of the H7N1 subtype exhibiting severe conjunctivitis and swelling of the infraorbital sinuses

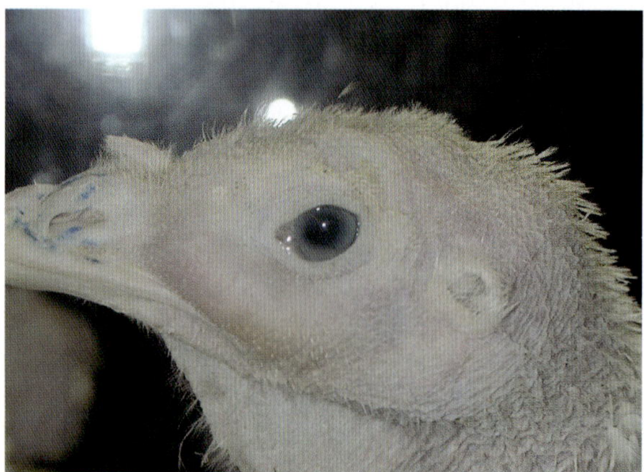

Fig. 6.3 Turkey experimentally infected with LPAI H7N3, showing conjunctivitis and swelling of the infraorbital sinuses

Fig. 6.4 28-day-old poults naturally infected with LPAI of the H7N1 subtype, showing severe depression. Subcutaneous emphysema results in the presence of air underneath the skin of the head

rapidly; soon after, respiratory signs appear. These start with respiratory distress, which initially features rales and snicking and then develops into severe dyspnoea, associated with swelling of the infraorbital sinuses (Fig. 6.2), and conjunctivitis (Fig 6.3). In some cases, the respiratory condition may become so severe that it results in air sac rupture and subcutaneous emphysema (Fig. 6.4).

In young birds, greenish or yellowish diarrhoea with the presence of undigested material in the faeces may occur. This condition is generally associated with higher mortality, greater difficulties in recovery, and poor weight gain.

Sexually mature turkey breeders often display milder forms of the respiratory complex, clinically characterised by rales, coughing, facial oedema and swelling of the infraorbital sinuses. The respiratory condition is usually accompanied by depression, reluctance to move and a febrile condition associated with loss of appetite. In most cases, the reproductive system is seriously compromised by viral replication. Losses in egg production may be severe during the acute phase. Egg quality may also be affected, with misshapen, fragile and whitish eggs being laid during the egg-drop phase (Fig. 6.5a,b).

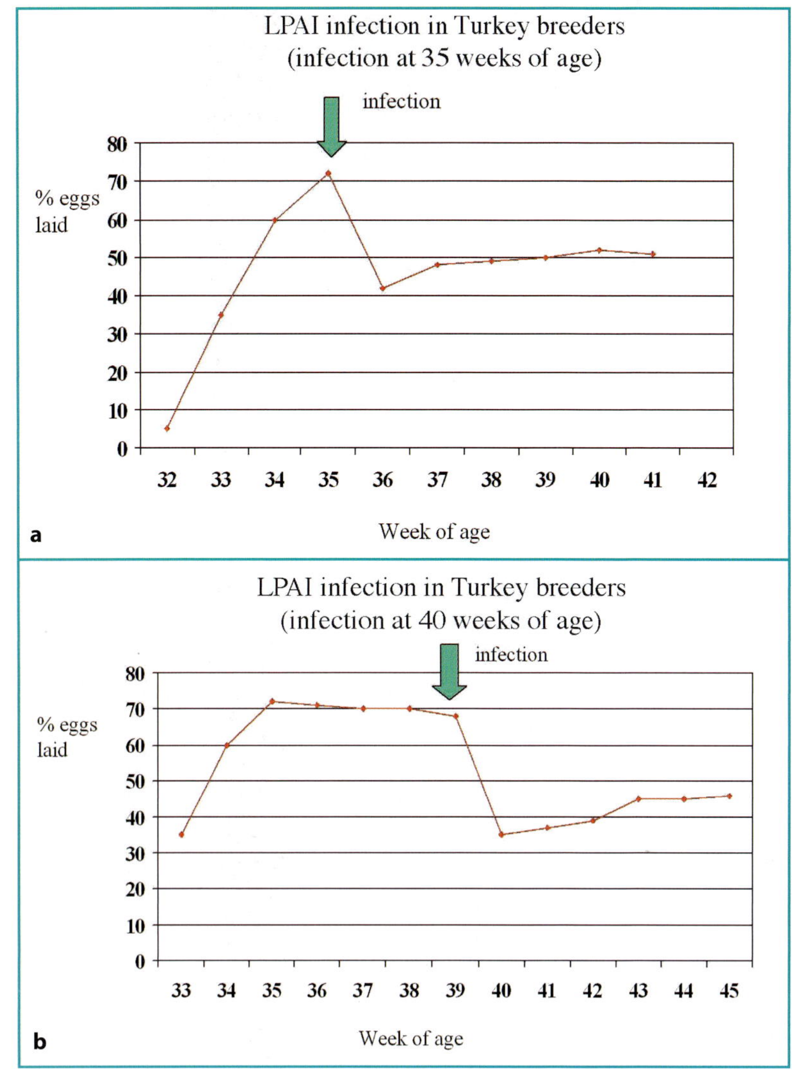

Fig. 6.5 a, b Turkey breeder egg-laying curve at 36 and 40 weeks of age following infection with LPAI

6.2.1.2 Gross Lesions

Both in adult and in young turkeys, a common post-mortem lesion is the presence of a caseous clot in the sinuses (Fig. 6.6) and trachea, which may cause death by suffocation. The trachea and lungs are oedematous, congested and in some cases haemorrhagic (Fig. 6.7). In adult birds a fibrinous air sacculitis involving the thoracic and abdominal air sacs and pericardium (Fig. 6.8) may result from secondary bacterial infections. The spleen is often enlarged and congested (Fig. 6.9).

The pancreas is generally affected in both adult and young birds. However, in young birds the changes are often more serious, as the pancreas may be enlarged and hardened, and may exhibit haemorrhages and necrotic areas on its surface (Fig. 6.10). In adult birds it appears congested, with pinpoint haemorrhages on its surface; this may be accompanied by congestion of the duodenal loop (Fig. 6.11). In convalescent birds the pancreas is reduced in size and atrophic. Petechial haemorrhages may be found on the epicardium and on the caecal tonsils (Fig. 6.12).

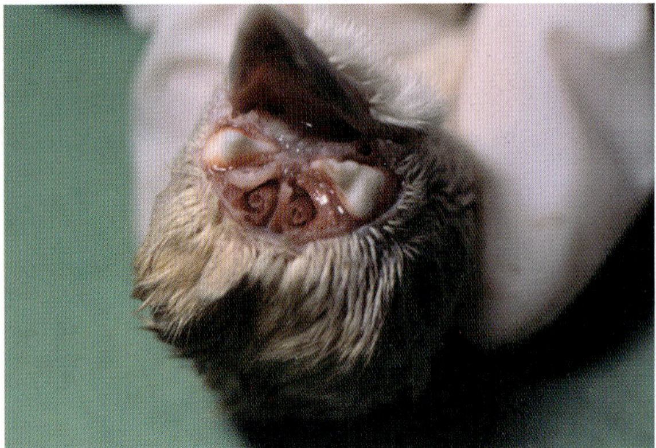

Fig. 6.6 28-day-old poult naturally infected with LPAI of the H7N1 subtype, presenting fibrin clots in the infraorbital sinuses

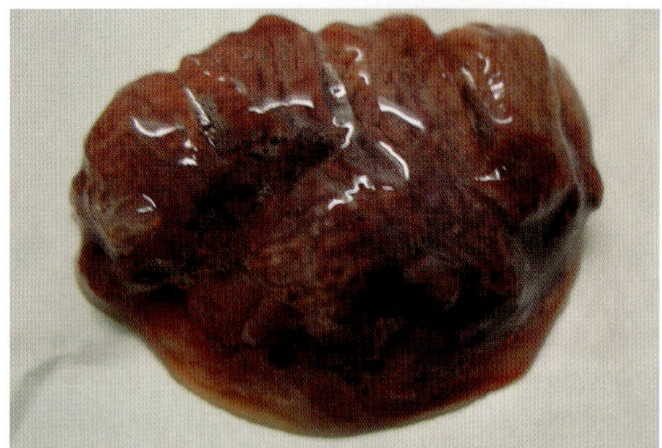

Fig. 6.7 Turkey experimentally infected with LPAI of the H7N1 subtype, exhibiting pneumonia with oedema

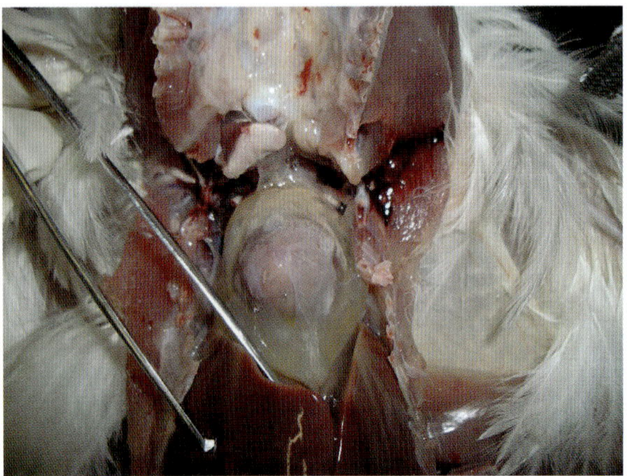

Fig. 6.8 Turkey experimentally infected with LPAI of the H7N1 subtype, displaying severe pericarditis with abundant serous exudates

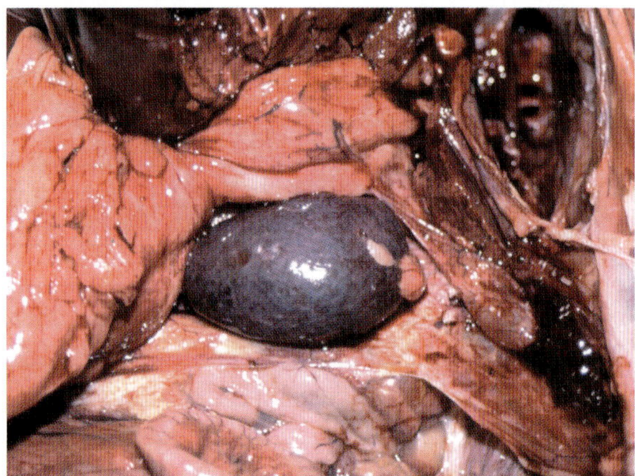

Fig. 6.9 Adult meat turkey naturally infected with LPAI of the H7N1 subtype, displaying splenomegaly

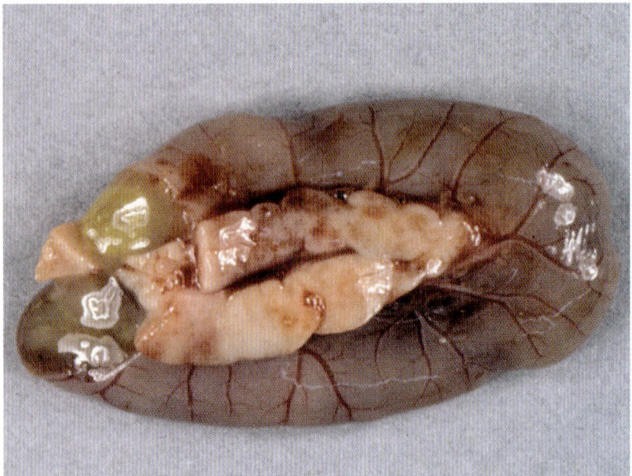

Fig. 6.10 28-day-old turkeys naturally infected with LPAI of the H7N1 subtype, displaying pancreatitis and duodenal distension

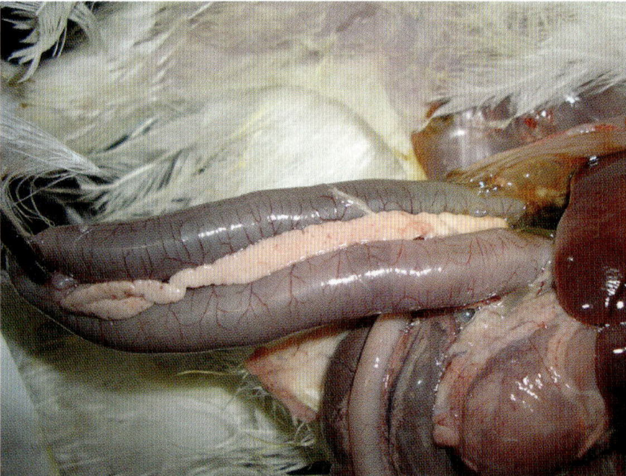

Fig. 6.11 Turkey experimentally infected with LPAI of the H7N3 subtype, displaying pancreatitis and duodenal distension

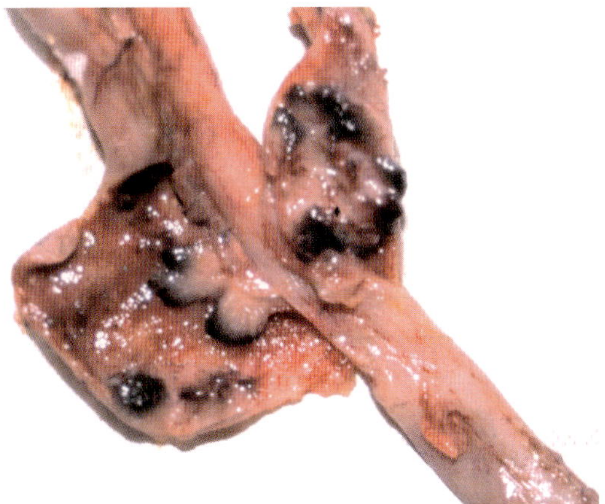

Fig. 6.12 Adult meat turkey naturally infected with LPAI of the H7N1 subtype, displaying haemorrhages on the caecal tonsils

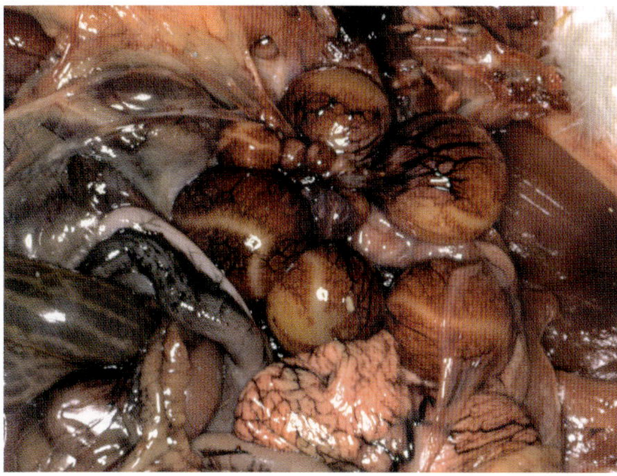

Fig. 6.13 Turkey breeder naturally infected with LPAI of the H7N1 subtype, displaying congestion of the ovary and egg-yolk peritonitis

On post-mortem examination, turkey breeders exhibit lesions primarily in the respiratory and reproductive tracts. Congestion of the lungs and trachea as well as sinusitis and conjunctivitis are common post-mortem findings and are often associated with "egg-yolk peritonitis". The latter arises from congestion of the ovary and haemorrhages of the ovarian follicles as well as congestion and oedema of the oviduct, which often contains a catarrhal-caseous exudate. Egg-yolk may also be found loose in the abdominal cavity (Fig. 6.13).

6.2.2 Chickens (*Gallus gallus*)

6.2.2.1 Clinical Signs

Chickens are less susceptible than turkeys from a clinical point of view. In broiler chickens, LPAI infections are often inapparent and, when present, can be confused with other conditions. When clinically overt, signs of infection are anorexia and mild respiratory distress, with flock mortality in the order of 2–3%. Among the clinical signs, rales, snicking and light coughing, in rare cases associated with conjunctivitis, may be observed. However, some LPAI viruses have been reported to cause serious health problems in broilers. This appears to be especially true of viruses of the H9N2 subtype, and these have been isolated from infections associated with significant mortality (sometimes reaching 50%) in the Middle East. There is field and laboratory evidence that in some of these cases there was an underlying bacterial or viral infection. Clinical signs noted in H9N2 outbreaks include swelling of the periorbital tissues and sinuses, respiratory discharge and severe respiratory distress (Fig. 6.14a,b).

Fig. 6.14 a, b Broiler breeder naturally infected with H9N2 LPAI, displaying severe periorbital oedema. (Courtesy of Nadim Mukhles Amarin)

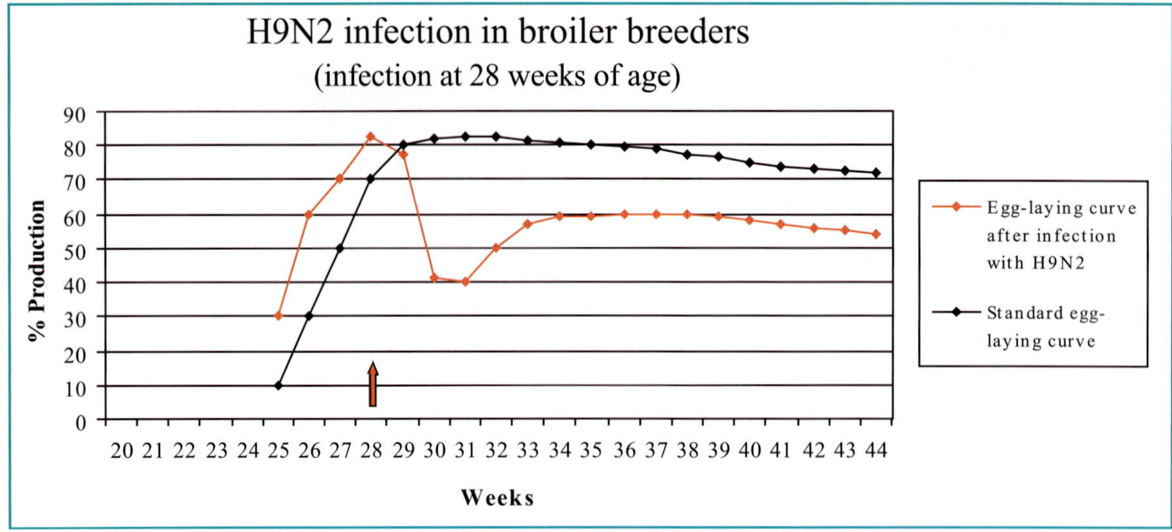

Fig. 6.15 Broiler breeder egg-laying curve following natural infection with H9N2 LPAI virus (*red line*). (Courtesy of Nadim Mukhles Amarin)

Sexually mature broiler breeders may be affected by a febrile condition accompanied by depression, somnolence and loss of appetite, all of which can be followed by a drop in egg production (Fig. 6.15). In this phase, cyanosis of the combs and wattles may be seen in a limited number of birds, accompanied by mild respiratory signs. During acute illness, eggs that are misshapen and/or with loss of colour may be laid in significant quantities.

6.2.2.2 Gross Lesions

Generally, broilers and young breeders do not exhibit any relevant post-mortem lesions. A very mild and restricted pulmonary and tracheal congestion may occur, together with catarrhal tracheitis.

During H9N2 outbreaks, gross lesions include extensive hyperaemia of the respiratory system followed by exudation and cast formation at the tracheal bifurcation, extending into the secondary bronchi.

In broiler breeders, pathological findings are restricted primarily to the ovary and oviduct. Ovarian follicles often appear haemorrhagic, oedematous and colliquated. The oviduct may be oedematous, with catarrhal or fibrinous egg-yolk peritonitis (Fig. 6.16). The only other consistent lesion is congestion of the lungs and trachea, which in some cases is associated with pulmonary oedema. Occasionally, pinpoint haemorrhages on the epicardium, liver and intestinal serosa are noted.

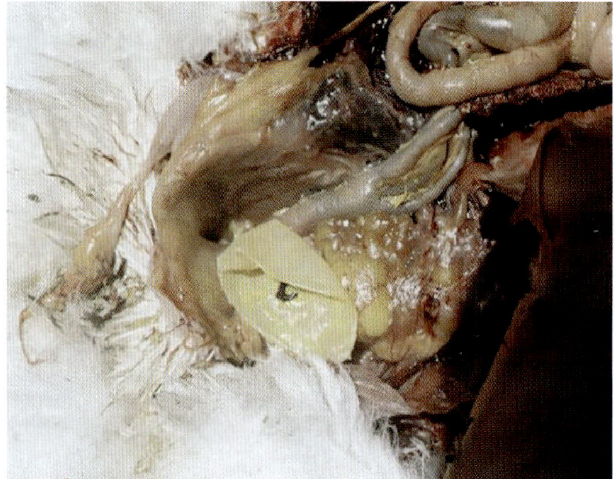

Fig. 6.16 Broiler breeder naturally infected with LPAI of the H9N2 subtype, displaying egg-yolk peritonitis. (Courtesy of Nadim Mukhles Amarin)

In layers, gross lesions mainly involve the reproductive organs: the ovary and oviduct appear oedematous and haemorrhagic, and the oviduct may contain a catarrhal exudate and caseous clots, often associated with secondary egg-yolk peritonitis (Fig. 6.17). The lungs and trachea are sometimes congested. Mild pancreatitis is occasionally seen.

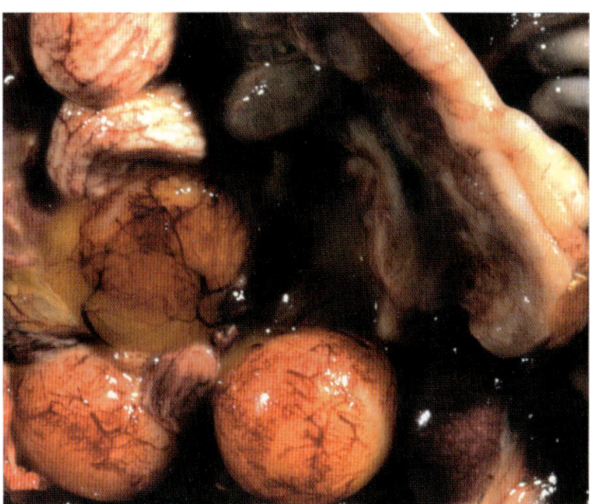

Fig. 6.17 Commercial egg layers naturally infected with LPAI of the H7N1 subtype, exhibiting ovarian congestion and egg-yolk peritonitis

6.2.3 Ostriches (*Struthio camelus*)

The first reported outbreaks of LPAI from ratites were caused by viruses of H7N1 subtype. During this epizootic in ostriches in South Africa in 1991, high mortality was seen in young birds.

6.2.3.1 Clinical Signs

Clinical signs of LPAI in ostriches are more evident in young birds (< 8 months of age) and include prostration, reluctance to feed and drink and enteric signs, typically greenish-white droppings. Other signs are ocular discharge and respiratory distress.

Various management and environmental factors contribute to the severity of the disease, such as poor hygiene, changing of feed composition, cold weather and the presence of secondary pathogens. After 2–3 days from the onset of symptoms affected birds can die or may gradually recover. In some cases, mortality is as high as 30%.

When affected clinically, ostriches > 14 months of age develop a very mild form of disease, with slight depression, reduction of food consumption and green discolouration of the urine. Adults may exhibit no clinical signs at all.

6.2.3.2 Gross Lesions

Post-mortem lesions are of a different nature and intensity, probably as a result of concurrent infections and complicating external factors. The most striking lesions are found in the liver, which may have a mottled appearance (due to multifocal areas of necrosis) and is swollen, congested and friable. In young ostriches, enteric lesions such as congestion of the cranial part of the small intestine and catarrhal enteritis, fibrinous air sacculitis and mucoid sinusitis can be present. Foci of necrosis and haemorrhages may be seen in the pancreas. Lesions caused by secondary bacterial infections (*E. coli*, *Pseudomonas aeruginosa*, *Staphylococcus aureus*, *Aspergillus fumigatus*) are very common.

6.2.4 Guinea Fowl (*Numida meleagris*)

6.2.4.1 Clinical Signs

In Guinea fowl, LPAI has been reported only rarely. In breeders, it resembles the clinical condition already described for turkey breeders. Depression and respiratory distress, characterised by rales and snicking, are preceded by a drop in egg production. Conjunctivitis appears to be a severe, constant clinical sign and is often complicated by secondary bacterial infections.

LPAI infection causes respiratory signs in Guinea fowl broilers similar to those observed in meat turkeys, with facial oedema being a prominent feature.

6.2.4.2 Gross Lesions

On post-mortem examination, congestion of internal organs and egg-yolk peritonitis have been reported in breeders. Pancreatitis, duodenitis and air sacculitis have been described both in breeders and in Guinea fowl broilers.

6.2.5 Japanese Quail (*Coturnix coturnix japonica*)

In this species, LPAI infection causes mild respiratory signs, including conjunctivitis, sneezing and clear nasal discharge. This may be associated with a complete loss of appetite.

In general, there are no significant post-mortem findings apart from congestion of the respiratory tract and pancreas.

6.2.6 Ducks and Geese

Ducks and geese are considered to be the natural reservoirs of LPAI viruses and are not known to exhibit clinical or pathological lesions following infection.

6.3 Highly Pathogenic Avian Influenza in Poultry

6.3.1 Turkeys (*Meleagris gallopavo*)

6.3.1.1 Clinical Signs

Clinical signs reflect viral replication and damage to multiple visceral organs as well as to the cardiovascular and nervous systems. In most cases the disease is fulminant, with some birds found dead prior to the observance of any clinical signs.

In meat and breeder turkeys, 100% flock mortality may be observed 48–72 h from the onset of the first clinical signs (Fig. 6.18a). Usually, there is a sudden and dramatic drop in food consumption, associated with depression and somnolence (Fig. 6.18b); this is followed by nervous signs, mainly tremors and incoordination. The birds exhibit shaking of the head, paralysis of the wings, abnormal gait and are often unable to maintain an upright position. They may lose their balance and end up in a recumbent position, making pedalling movements with their legs. Spastic contraction of the wings, associated with flapping movements and opistothonus, are also seen. Due to the dramatic and acute nature of the nervous signs, affected birds may be found dead lying on their backs (Fig. 6.18c).

6.3.1.2 Gross Lesions

On post-mortem examination, the internal organs are often congested but in some birds no post-mortem findings are detected, due to the peracute nature of the disease. Pancreatitis (Fig. 6.19) is typically present in ad-

Fig. 6.18 a-c Turkey breeder naturally infected with HPAI of the H7N1 subtype, which causes mass mortality 72 h from the onset of the first clinical signs (**a**). Paralysis of the wings and severe depression (**b**) are evident. The birds are often found dead, lying on their backs due to nervous signs and spastic contractions prior to death (**c**)

Fig. 6.19 Meat turkey naturally infected with HPAI of the H7N1 subtype, exhibiting haemorrhagic pancreatitis and duodenal distension

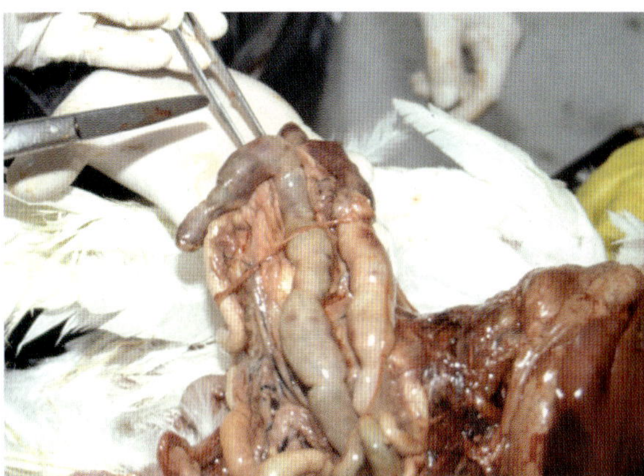

Fig. 6.20 Turkey breeder naturally infected with HPAI of the H7N1 subtype, displaying gaseous distension of caeca and serosal haemorrhages

Fig. 6.21 Adult meat turkey naturally infected with HPAI of the H7N1 subtype, exhibiting congestion and necrosis of the spleen

Fig. 6.22 Broiler breeders naturally infected with HPAI of the H5N1 subtype. Infection results in sudden high mortality, with birds found dead in nest boxes. (Courtesy of Ahmed Abd El Karim)

dition to haemorrhages on serosal or mucosal surfaces and foci of necrosis within parenchyma of visceral organs. In particular, haemorrhages are found on the mucosa of the proventriculus and ventriculus and caecal tonsils (Fig. 6.20). On occasion, the spleen is enlarged and congested (Fig. 6.21) and pinpoint haemorrhages are seen on the epicardium. In a limited number of cases, kidney congestion and urate deposits are present.

6.3.2 Chickens (*Gallus gallus*)

6.3.2.1 Clinical Signs

Chickens are also highly susceptible to infection and clinical disease. Moreover, management systems may influence the transmission dynamics within the flock and to a certain extent the clinical manifestations.

In chickens reared on litter, the transmission rate of the disease is very fast and flock mortality may be as high as 100% within 3–4 days from the onset of the first clinical signs (Figs. 6.22, 6.23). Anorexia, depression and cessation of egg-laying in breeders are followed by nervous signs, characterised by prostration, complete reluctance to move, tremors of the head, paralysis of the wings and incoordination of leg movements when the birds are stimulated to move (Figs. 6.24, 6.25). Typical lesions include cyanosis of the comb and wattles (Figs. 6.26–6.28) and petechial haemorrhages on the hock (Fig. 6.29), although in some outbreaks such signs have affected only a limited number of birds. Sudden death occurs in a recumbent position and is preceded by pedalling movements and gasping.

Individual birds that survive longer may exhibit nervous disorders, such as tremors of the head and

Fig. 6.23 Broiler breeders following natural infection with HPAI of the H5N1 subtype, displaying mass mortality a few days after the onset of clinical signs. (Courtesy of Ahmed Abd El Karim)

Fig. 6.24 Broiler breeders following natural infection with HPAI of the H5N1 subtype, displaying neurological signs. (Courtesy of Ahmed Abd El Karim)

Fig. 6.25 Caged layers naturally infected with HPAI of the H7N1 subtype, showing prostration and reluctance to move in the preagonic phase

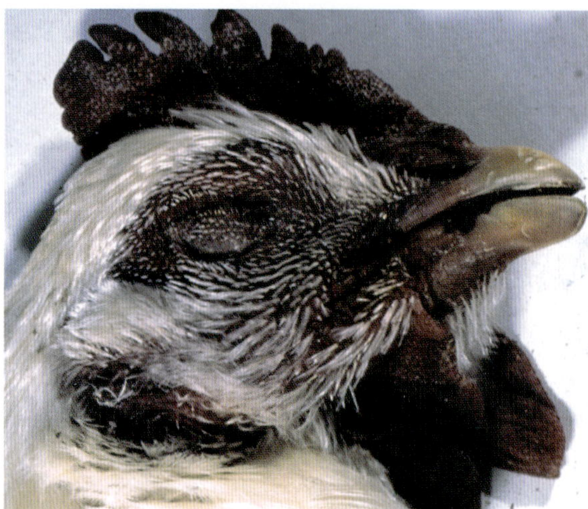

Fig. 6.26 Broiler breeders naturally infected with HPAI of the H7N1 subtype, displaying congestion and cyanosis of the comb and wattles

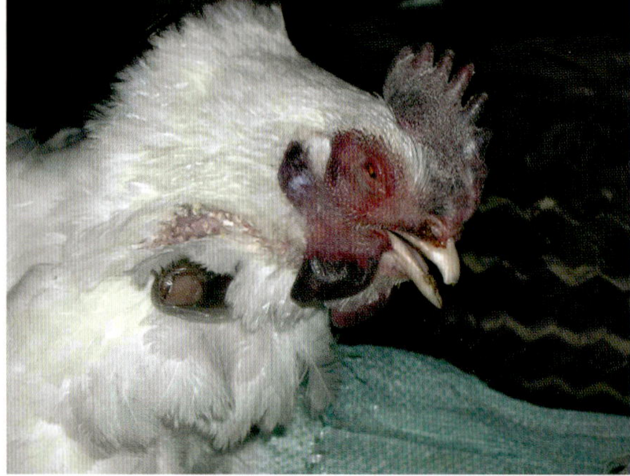

Fig. 6.27 Broiler breeders naturally infected with HPAI of the H5N1 subtype, showing severe congestion and oedema of the comb and wattles. (Courtesy of Walid Hamdy)

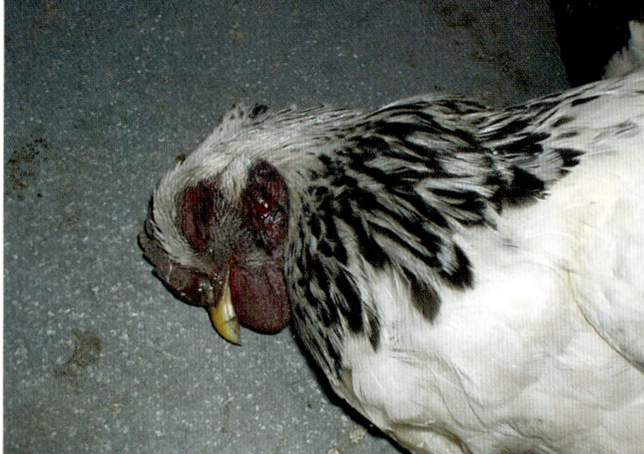

Fig. 6.28 Backyard chicken naturally infected with HPAI of the H5N1 subtype, showing necrosis of the comb and wattles and congestion of the unfeathered skin of the head. (Courtesy of Vladimir Savic)

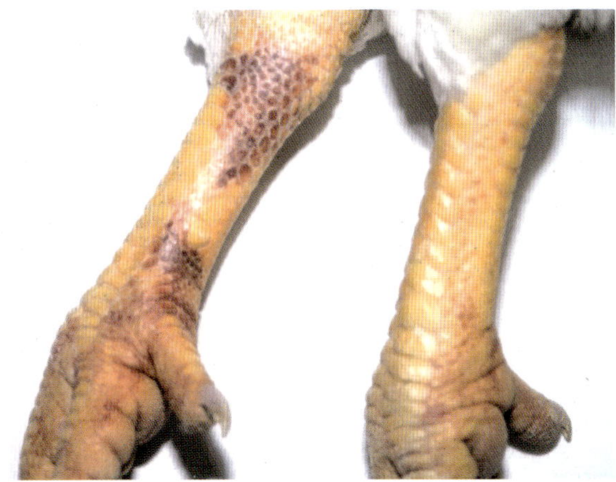

Fig. 6.29 Broiler breeders naturally infected with HPAI of the H7N1 subtype, showing haemorrhages on the shank

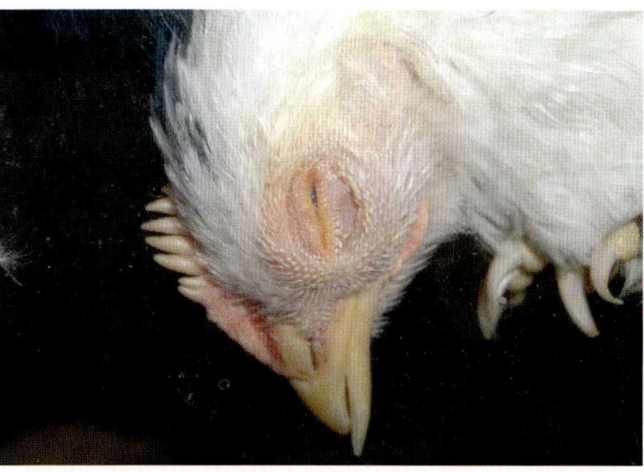

Fig. 6.30 Specific-pathogen-free (SPF) chicken experimentally infected with HPAI of the H5N1 subtype, exhibiting periorbital oedema

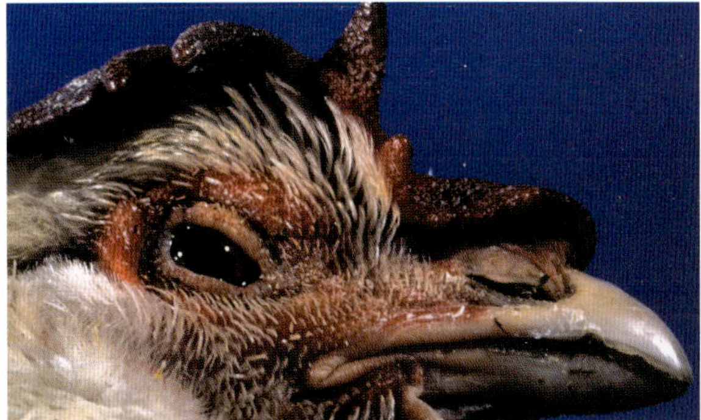

Fig. 6.31 Broiler breeder naturally infected with HPAI H7N1, exhibiting congestion of the comb and facial skin during the recovery phase

Fig. 6.32 Caged layers naturally infected with HPAI of the H7N1 subtype. Note the appearance of the poultry house following the onset of clinical signs

neck, inability to stand, torticollis, opisthotonus and abnormal gait. In these birds, there is severe congestion of the comb, conjunctivitis and periorbital oedema (Figs. 6.30, 6.31), ruffled feathers and severe depression. Recovery very rarely occurs.

In caged layers, the clinical manifestations may appear to affect only single birds initially, spreading to the rest of the flock very slowly. Early signs include severe depression or mortality in only one bird per cage in a restricted area of the house (Fig. 6.32); spreading subsequently to neighbouring cages, and affecting the whole shed within 10–14 days. This difference in spread between caged and litter-reared chickens is probably related to the reduced bird-to-bird contact occurring between cages and the absence of mingling between susceptible and infected animals. In addition, in caged-reared birds, the amount of infected faeces in direct contact with the birds is lower than for floor-raised birds. During the 2000 Italian epidemic of HPAI H7N1, the incubation period in one caged layer flock was estimated to be 18 days.

In layers, clinical signs consist of prostration, somnolence and cessation of egg laying and feed consumption. The combs appear cyanotic and in some cases pale and flaccid (Figs. 6.33). Haemorrhages on the hock, gasping associated with tremors of the head and prostration may also be seen. Birds released from their cages are immobile and discharge a serous greenish-yellow liquid from the oral cavity. Eggshell quality is also affected (whitish and fragile eggs),

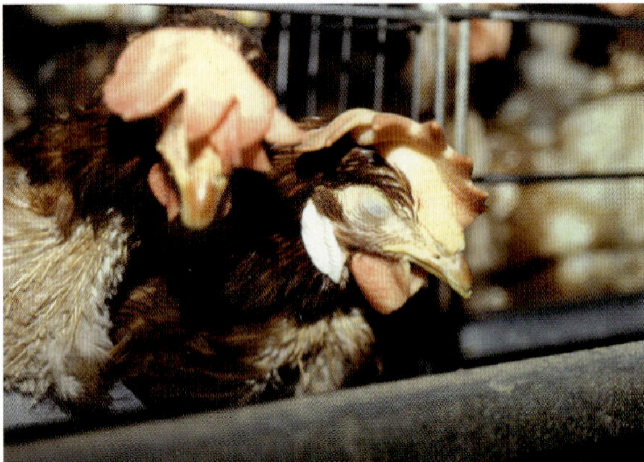

Fig. 6.33 Caged layers naturally infected with HPAI of the H7N1 subtype, exhibiting severe depression in the acute phase. Note the pale and flaccid combs

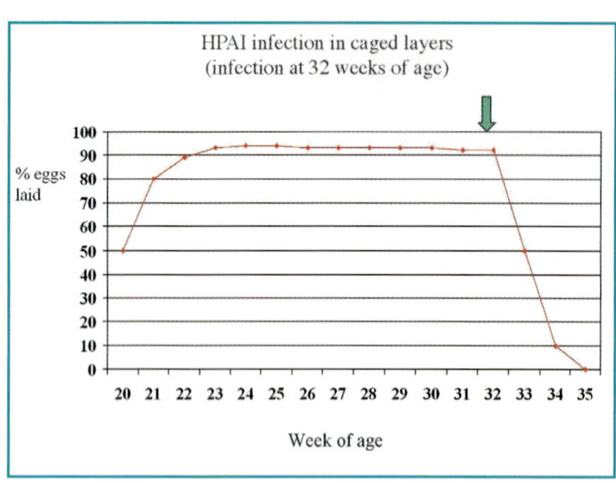

Fig. 6.34 Egg-laying curve of caged 32-week-old layers naturally infected with HPAI of the H7N1 subtype

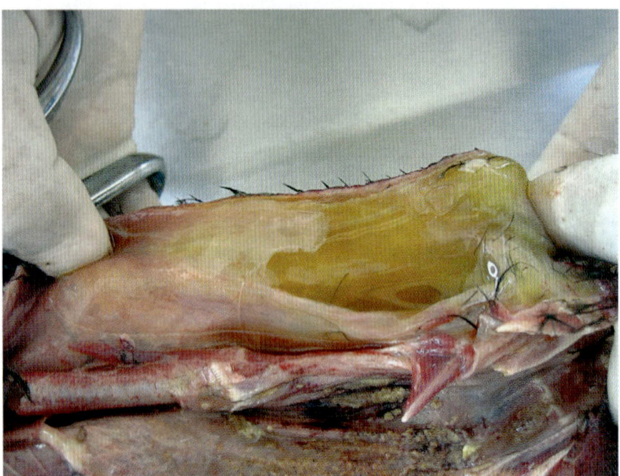

Fig. 6.35 Chicken naturally infected with HPAI of the H5N1 subtype, presenting with abundant serofibrinous subcutaneous oedema in tissues surrounding the neck. (Courtesy of Thierry Van den Berg)

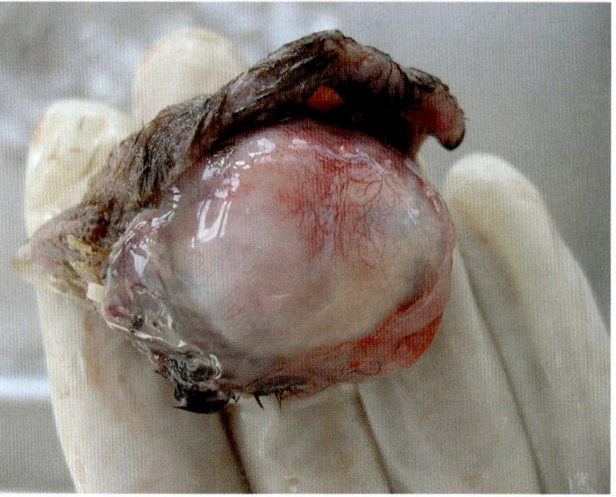

Fig. 6.36 Chicken naturally infected with HPAI of the H5N1 subtype, showing congestion of the subcutaneous tissue of the head. (Courtesy of Thierry Van den Berg)

probably also due to the birds' complete loss of appetite. The number of eggs produced decrease very rapidly, due to the high mortality (Fig. 6.34).

6.3.2.2 Gross Lesions

Swelling of the head and upper neck are common as a result of subcutaneous oedema and may be accompanied by petechial or ecchymotic haemorrhages (Figs. 6.35, 6.36.). Conjunctivitis and periorbital oedema often occur. Haemorrhages and cyanosis of the skin are frequent, especially in the wattles, combs and legs. Lesions in visceral organs are represented by haemorrhages on serosal or mucosal surfaces and abdominal fat (Fig. 6.37, 6.38a, 6.38b, 6.39,) and foci of necrosis (Fig. 6.40) within the parenchyma. Haemorrhages on the epicardium (Fig. 6.41), pericardium (Fig. 6.42), in pectoral and leg (Figs. 6.43, 6.44) muscles, in the serosa and mucosa of the proventriculus (Fig. 6.45), ventriculus and caecal tonsils are common (Fig. 6.46). The organ that appears to be damaged most consistently by viral replication is the pancreas (Figs. 6.47, 6.48), which exhibits focal to diffuse necrosis of the acinar cells. Occasionally, interstitial oedema of the pancreas associated with fibrinous peritonitis is seen. The lungs and trachea are congested (Fig. 6.49); the latter exhibits haemorrhages of the submucosa.

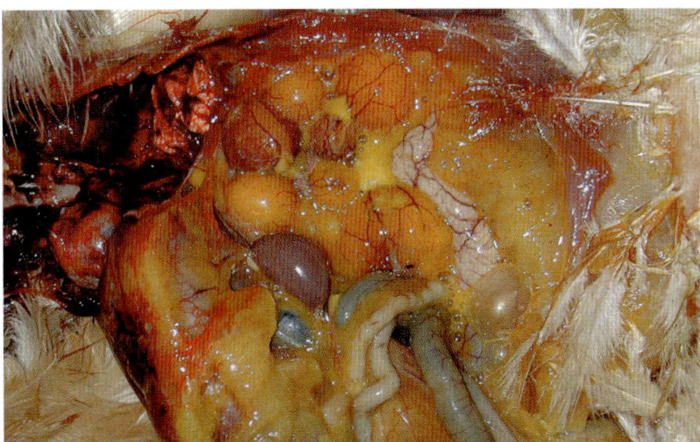

Fig. 6.37 Layers naturally infected with HPAI of the H7N1 subtype, displaying egg-yolk peritonitis

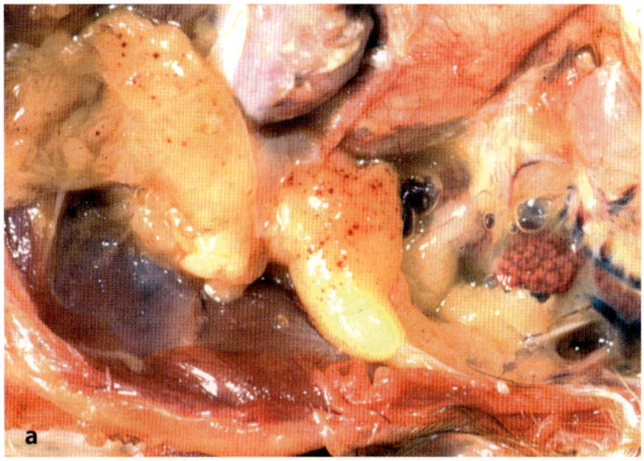

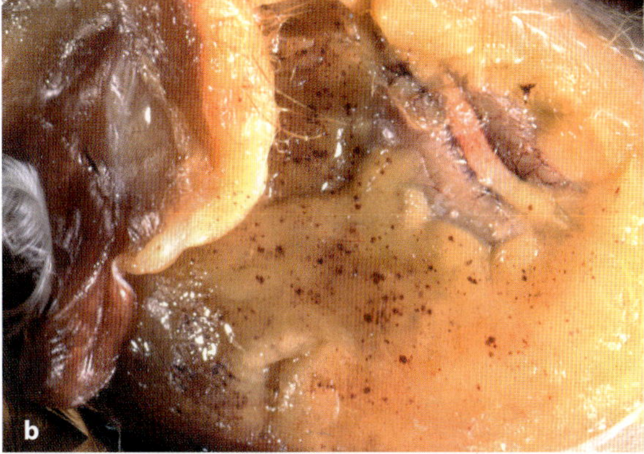

Fig. 6.38 a, b Caged layers naturally infected with HPAI of the H7N1 subtype, displaying haemorrhages on the abdominal fat

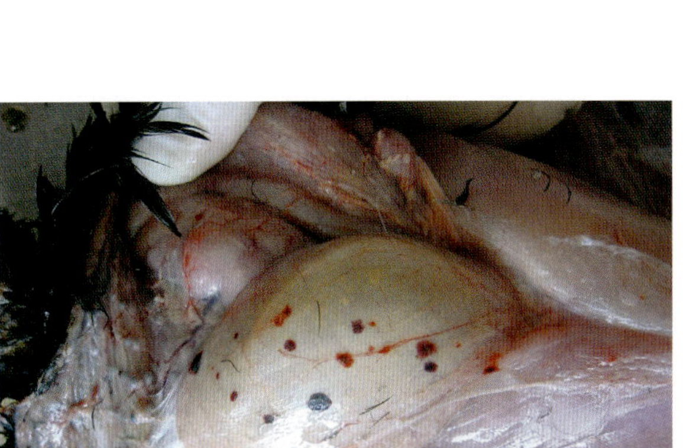

Fig. 6.39 Chicken naturally infected with HPAI of the H5N1 subtype, showing petecchial haemorrhages on the serosal surface of the crop. (Courtesy of Thierry Van den Berg)

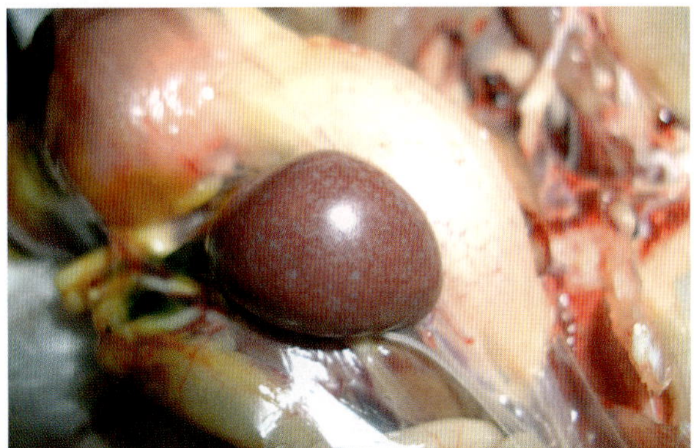

Fig. 6.40 SPF chicken experimentally infected with HPAI of the H7N1 subtype, showing hyperplasia and necrosis of the spleen

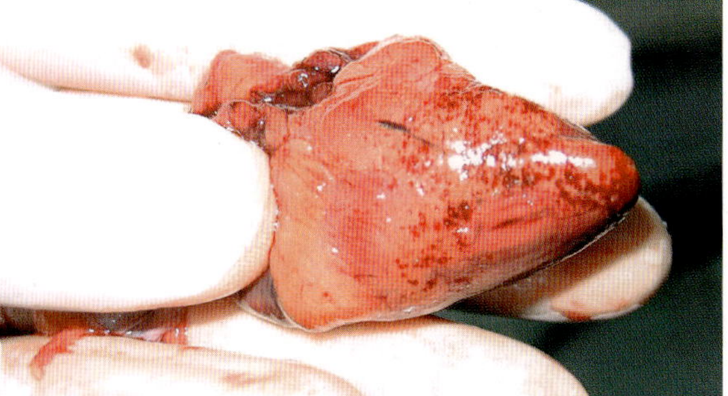

Fig. 6.41 Chicken naturally infected with HPAI of the H5N1 subtype, displaying petecchial haemorrhages in the epicardium. (Courtesy of Thierry Van den Berg)

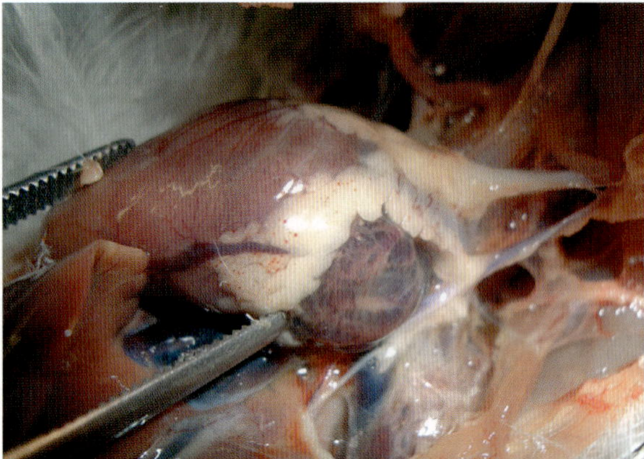

Fig. 6.42 SPF chicken experimentally infected with HPAI of the H7N1 subtype, exhibiting pericarditis with clear exudate and petecchial haemorrhages on pericardial fat

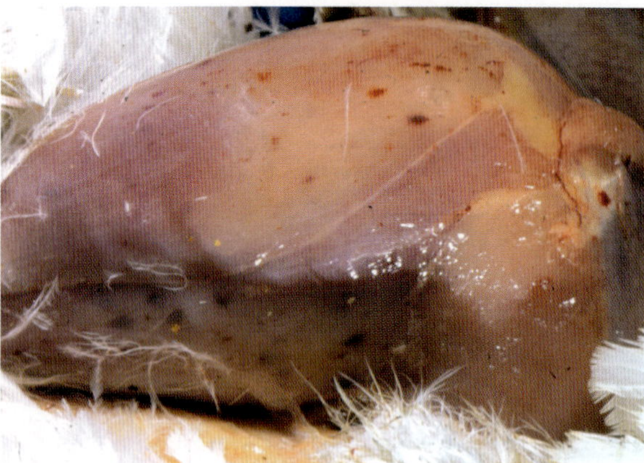

Fig. 6.43 Chicken naturally infected with HPAI of the H7N1 subtype, showing petecchial haemorrhages in leg muscles

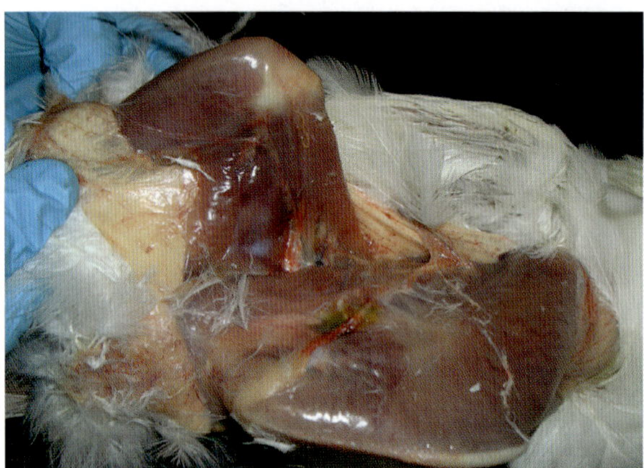

Fig. 6.44 SPF chicken naturally infected with HPAI of the H7N1 subtype, displaying stripe-shape haemorrhages in breast and leg muscles

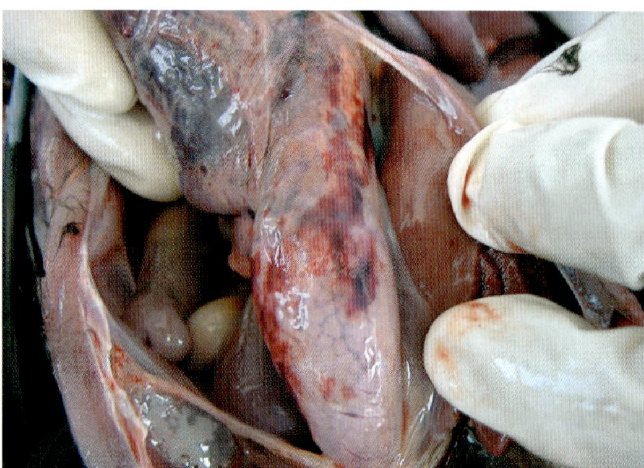

Fig. 6.45 Chicken naturally infected with HPAI of the H5N1 subtype, displaying haemorrhages on the serosal surface of the proventriculus. (Courtesy of Thierry Van den Berg)

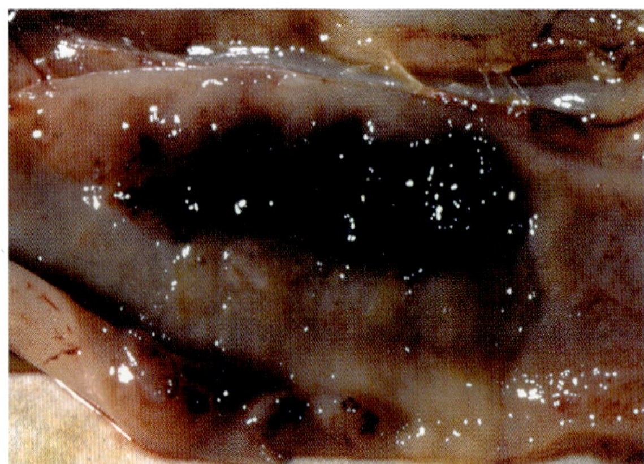

Fig. 6.46 Broiler breeders naturally infected with HPAI of the H7N1 subtype, displaying haemorrhages on the caecal tonsils

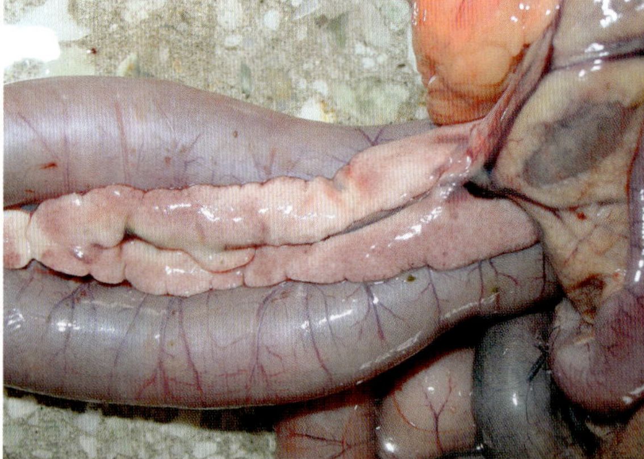

Fig. 6.47 Chicken naturally infected with HPAI of the H5N1 subtype, exhibiting haemorrhagic pancreatitis. (Courtesy of Corrie Brown)

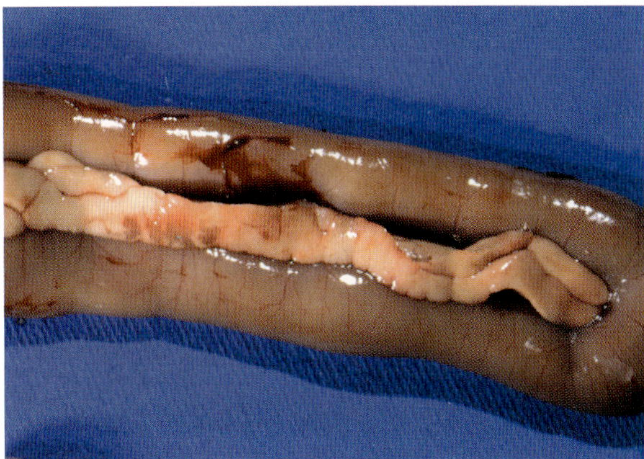

Fig. 6.48 Broiler breeders naturally infected with HPAI of the H7N1 subtype, displaying pancreatitis

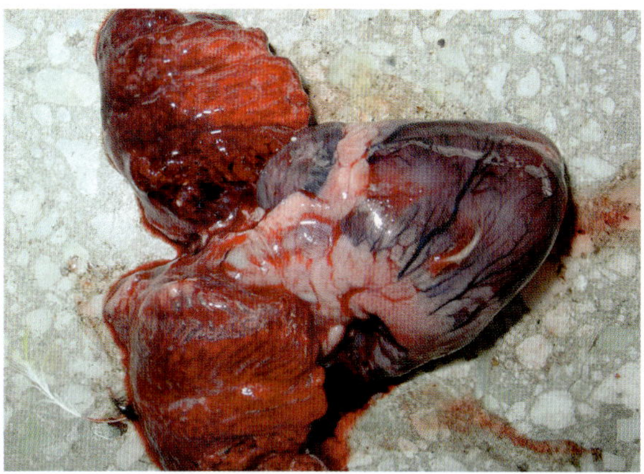

Fig. 6.49 Chicken naturally infected with HPAI of the H5N1 subtype, showing bilateral pneumonia with oedema. (Courtesy of Corrie Brown)

Fig. 6.50 Brilliant-green urine containing urate deposits produced by an ostrich naturally infected with HPAI of the H7N1 subtype

Fig. 6.51 Haemorrhagic faeces produced by an ostrich affected by HPAI of the H7N1 subtype

6.3.3 Ostriches (*Struthio camelus*)

6.3.3.1 Clinical Signs and Gross Lesions

Clinical signs and lesions in the visceral organs of ostriches infected with HPAI viruses vary depending on the viral strain, but occur most consistently in young or juvenile birds. Early signs include reduced activity and appetite prior to the appearance of clear clinical signs, such as depression, ruffled feathers, sneezing and open mouth breathing. Some birds show incoordination, paralysis of the wings and tremors of the head and neck. The head may appear congested and swollen and in some cases completely bent backwards or sideways as a result of torticollis.

The urine appears brilliant green (Fig. 6.50) and is rich in urates; the faeces are haemorrhagic (Fig. 6.51). In the preagonic phase, birds lie on the ground and discharge a green mucus liquid from the oral cavity (Fig. 6.52). Some birds may recover after approximately one week from the onset of the clinical signs. Adults or breeders generally do not show any clinical signs when infected with HPAI viruses.

Fig. 6.52 Juvenile ostrich affected by HPAI of the H7N1 subtype, exhibiting greenish liquid discharge from the nostrils

Fig. 6.53 Juvenile ostrich affected by HPAI of the H7N1 subtype, showing necrotic haemorragic enteritis

Fig. 6.54 Juvenile ostrich affected by HPAI of the H7N1 subtype, showing pinpoint haemorrhages on the intestinal serosa

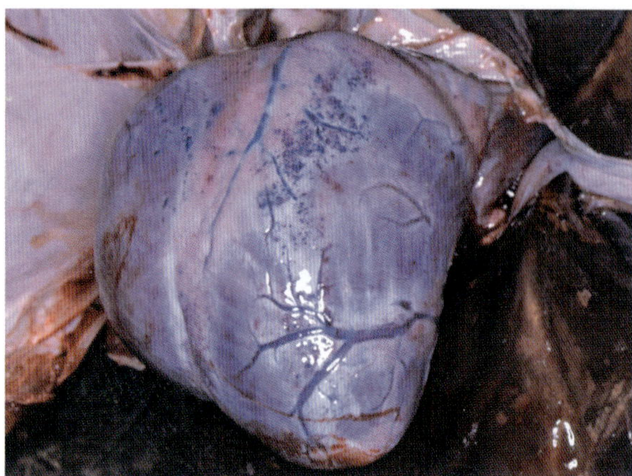

Fig. 6.55 Juvenile ostrich affected by HPAI of the H7N1 subtype, displaying pinpoint haemorrages on the epicardium

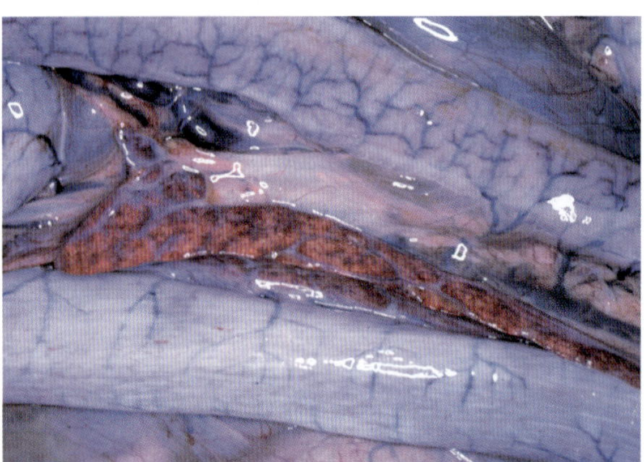

Fig. 6.56 Juvenile ostrich affected by HPAI of the H7N1 subtype, displaying necrotic haemorrhagic pancreatitis

On post-mortem examination the lesions vary with viral strain; however, common findings are oedema of the head and upper part of the neck, haemorrhagic enteritis with haemorrhagic exudate in the intestinal lumen (Fig. 6.53), pinpoint or petechial haemorrhages on the surface of the intestinal serosa (Fig. 6.54) and the epicardium (Fig. 6.55), a haemorrhagic, enlarged and hardened pancreas, (Fig. 6.56) as well as congestion in the lung and trachea. The liver may be enlarged, with rounded margins, and contain irregular areas of necrosis and congestion (Fig. 6.57). The kidneys are enlarged, softened and contain urate deposits. The spleen may also be larger.

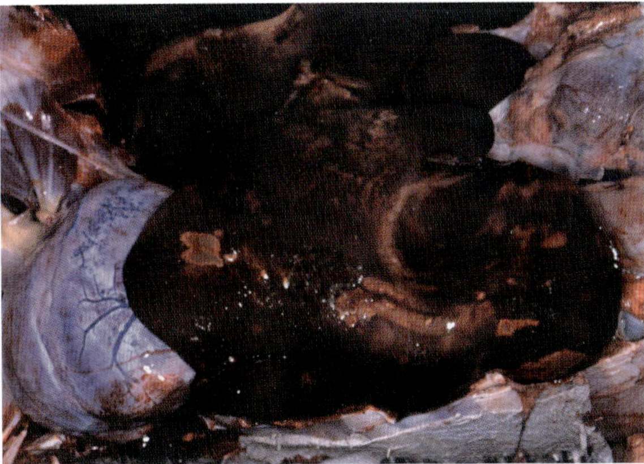

Fig. 6.57 Juvenile ostrich naturally infected with HPAI H7N1 subtype, showing multiple isolated-to-coalescent necrotic foci on the liver surface

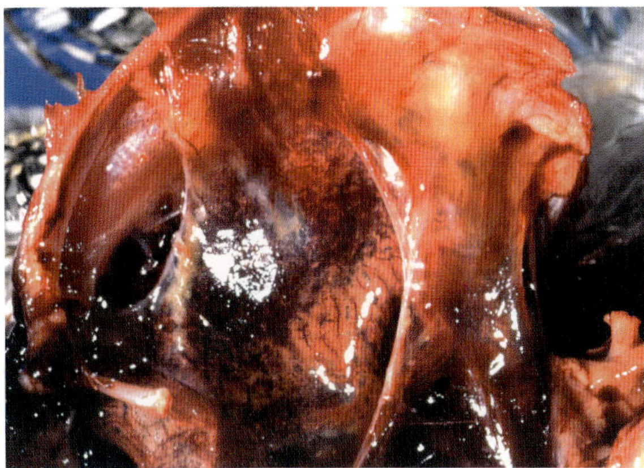

Fig. 6.58 Guinea-fowl naturally infected with HPAI of the H7N1 subtype, presenting pulmonary congestion and oedema

6.3.4 Guinea Fowl (*Numida meleagris*)

6.3.4.1 Clinical Signs

Very few cases of HPAI infections of Guinea fowl have been reported worldwide and even fewer have been documented. In general, these birds seem to be highly susceptible to HPAI. Sudden death syndrome, with very high mortality (up to 100% of breeding birds), has been reported to occur 48–72 h from the onset of the first clinical signs. Clinical signs include anorexia and depression, followed by nervous signs.

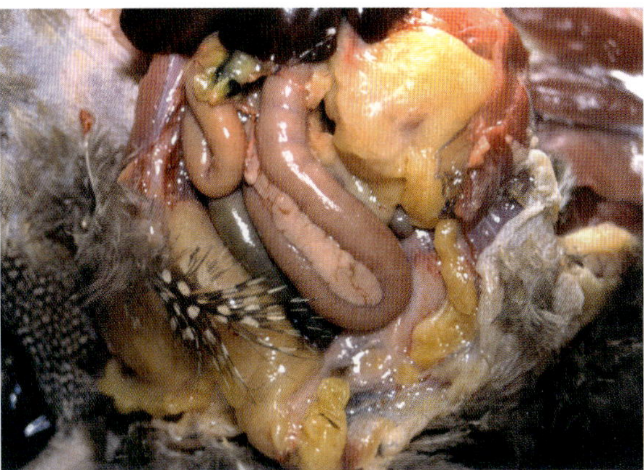

Fig. 6.59 Guinea-fowl naturally infected with HPAI of the H7N1 subtype, exhibiting pancreatitis

6.3.4.2 Gross Lesions

On post-mortem examination, the gross lesions are similar to those observed in chickens. A common finding is congestion of the internal organs, associated in some cases with pulmonary oedema (Fig. 6.58). Focal pancreatitis is also reported (Fig. 6.59).

6.3.5 Japanese Quail (*Coturnix coturnix japonica*)

6.3.5.1 Clinical Signs

In this species, HPAI infections are characterised by a severe respiratory condition that in a few days evolves into a clinical presentation comprising prostration, somnolence and listlessness, often accompanied by production of a whitish diarrhoea and gasping prior to death. Nervous signs such as opistothonus and torticollis (Fig. 6.60) are also commonly seen at the same time. Egg laying ceases within a few days of the onset of the first clinical signs. Mortality is lower than in chickens and turkeys, ranging between 5 and 10% per day.

6.3.5.2 Gross Lesions

On post-mortem examination, the only lesions that have been reported are congestion of the viscera (Figs. 6.61, 6.62), enteritis, egg-yolk peritonitis and haemorrhagic pancreatitis (Figs. 6.63, 6.64). Congestion of muscle has also been observed (Fig. 6.65). The peri-cloacal feathers are often soiled with greenish-white faecal material.

Fig. 6.60 Japanese quail (*Coturnix coturnix japonica*) experimentally infected with HPAI of the H7N1 subtype, showing torticollis

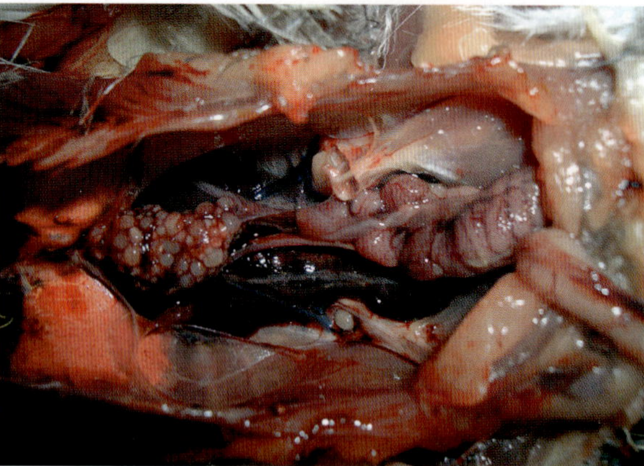

Fig. 6.61 Japanese quail (*Coturnix coturnix japonica*) experimentally infected with HPAI of the H7N1 subtype, exhibiting congestion in kidneys, ovary and oviduct

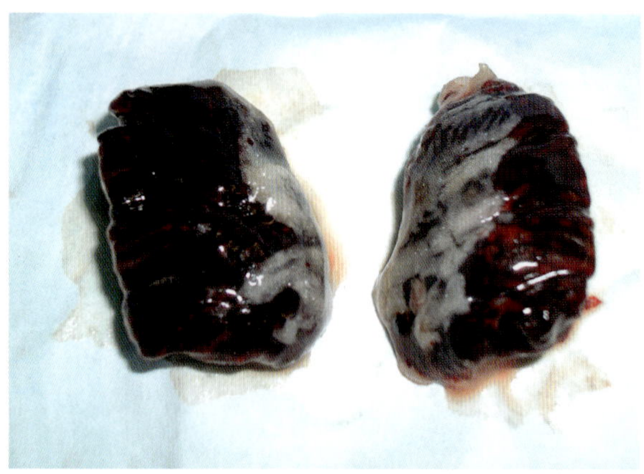

Fig. 6.62 Lungs of a Japanese quail (*Coturnix coturnix japonica*) experimentally infected with HPAI of the H7N1 subtype, showing bilateral pneumonia with congestion and oedema

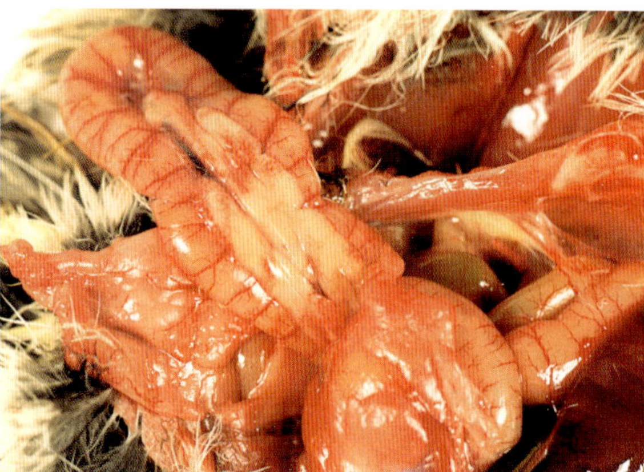

Fig. 6.63 Quail breeder naturally infected with HPAI of the H7N1 subtype, presenting with haemorrhagic pancreatitis

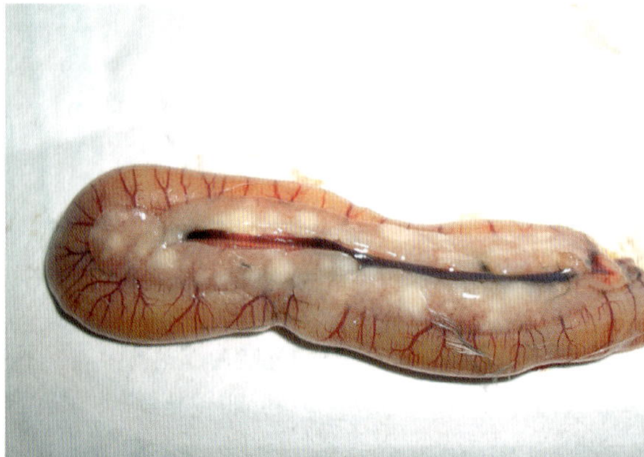

Fig. 6.64 Japanese quail (*Coturnix coturnix japonica*) experimentally infected with HPAI of the H7N1 subtype, showing pancreatitis with multifocal necrosis and duodenitis

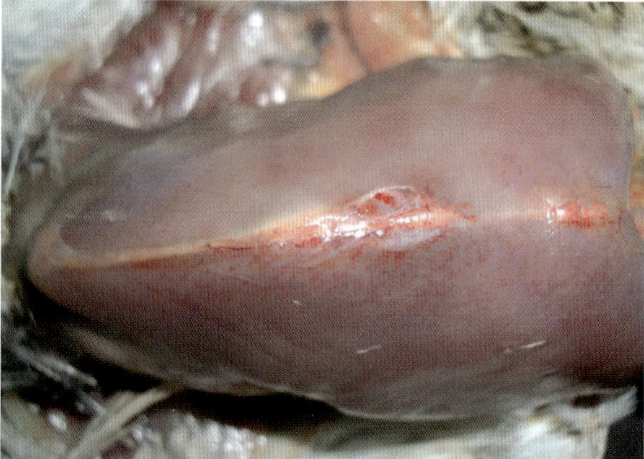

Fig. 6.65 Japanese quail (*Coturnix coturnix japonica*) experimentally infected with HPAI of the H7N1 subtype, exhibiting haemorrhagic breast muscles

Care should be taken in the clinical assessment of suspected HPAI in quail, as in experimental studies some strains of HPAI have failed to produce clinical signs.

6.3.6. Ducks and Geese

6.3.6.1 Clinical Signs

Ducks and geese are known to be more resistant to the clinical manifestations of HPAI than other domestic birds.

However, some strains, namely H7N1 and H5N1, have been shown to be pathogenic for several species of waterfowl, in some cases causing > 50% mortality. Muscovy ducks (*Cairina moschata*) have been reported to exhibit nervous signs such as incoordination and tremors and an abnormal gait similar to limping.

Pekin ducks (*Anas platyrhynchos*) are also considered to be resistant clinically to HPAI viruses. Reports from natural and experimental infection with viruses of the Asian H5N1 subtype indicated that birds infected with certain strains may not exhibit any clinical signs, whereas other strains lead to mortality of up to 70%. Clinical signs include conjunctivitis and mild depression, followed by nervous signs such as torticollis, incoordination, tremors and seizures (Figs. 6.66, 6.67).

Fig. 6.66 Pekin duck (*Anas platyrhinchos*) experimentally infected with HPAI of the H5N1 subtype, showing opisthotonus ("stargazing")

Fig. 6.67 Pekin duck (*Anas platyrhinchos*) naturally infected with HPAI of the H5N1 subtype, presenting torticollis. (Courtesy of Walid Hamdy Kilany)

6.3.6.2 Gross Lesions

Lesions in Pekin ducks infected with strains of HPAI viruses that produce clinical signs, detected on post-mortem examination, include widespread bleeding along the intestinal tract (Fig 6.68), especially the stomach and gizzard (Fig. 6.69). Haemorrhages on the surface of the pancreas and trachea and air sacculitis (Fig. 6.70) have been observed in some birds. Hyperaemia and pinpoint haemorrhages may be found on the surface of the brain (Fig. 6.71) and on the bill (Fig. 6.72). Naturally infected geese have been affected by pancreatitis (Fig. 6.73).

6.3.7 Wild Waterfowl

6.3.7.1 Clinical Signs

Across Europe, most of the wild waterfowl naturally infected with the HPAI H5N1 Asian virus have displayed neurological signs. The behavioural patterns of affected birds include a tendency to remain in isolation, reluctance to move when stimulated and absence of any sign indicating fear of human beings. Some birds were seen to have difficulty in surface

Fig. 6.68 Pekin duck (*Anas platyrhinchos*) experimentally infected with HPAI of the H5N1 subtype, presenting haemorrhagic content in the intestine

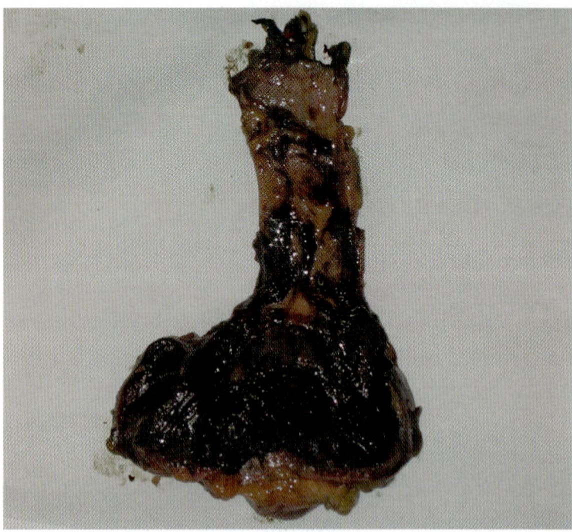

Fig. 6.69 Pekin duck (*Anas platyrhinchos*) experimentally infected with HPAI of the H5N1 subtype, exhibiting haemorrhagic content in the proventriculus and gizzard

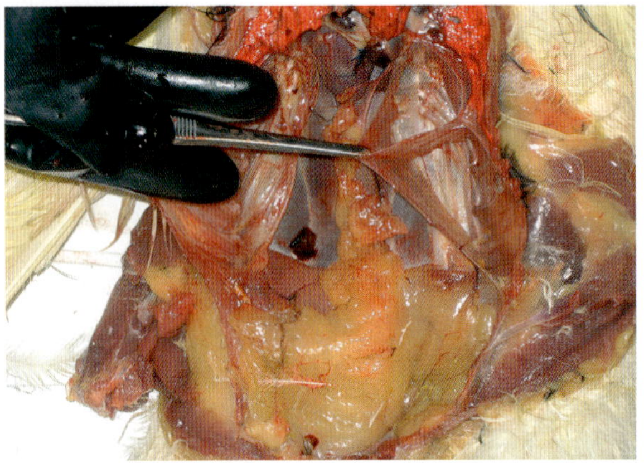

Fig. 6.70 Pekin duck (*Anas platyrhinchos*) experimentally infected with HPAI of the H5N1 subtype, exhibiting air sacculitis

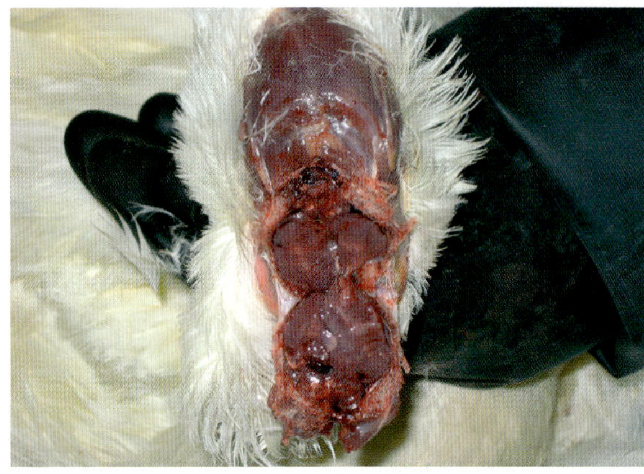

Fig. 6.71 Pekin duck (*Anas platyrhinchos*) experimentally infected with HPAI of the H5N1 subtype, displaying petecchial haemorrhages in the brain

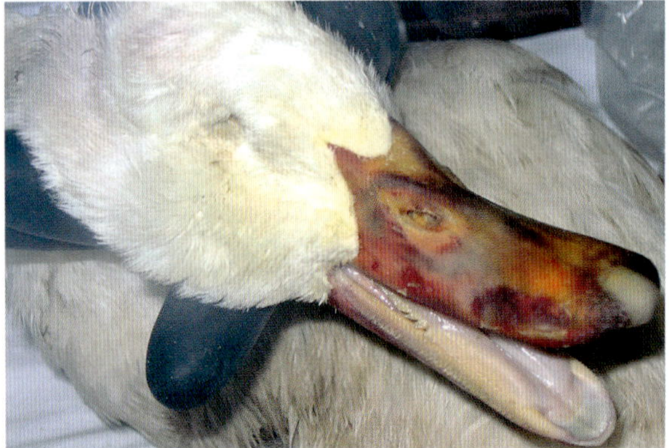

Fig. 6.72 Pekin duck (*Anas platyrhinchos*) experimentally infected with HPAI of the H5N1 subtype, showing haemorrhages on the bill

Fig. 6.73 Domestic goose naturally affected by HPAI of the H7N1 subtype, presenting pancreatitis and duodenitis

feeding and swimming and some moved in circles both on the ground and on water. Torticollis and tremors of the head and neck are a typical finding in clinically affected birds (Fig. 6.74).

6.3.7.2 Gross Lesions

Lesions are in-keeping with known findings for HPAI also in other avian species, i.e. haemorrhages and congestion of the pancreas (Figs. 6.75–6.80), congestion of the lungs and trachea, which may appear haemorrhagic (Figs. 6.81–6.83), air sacculitis, ecchymotic haemorrhages on the epicardium and cardiac muscle (Fig. 6.84), catarrhal or haemorrhagic enteritis and proventriculitis (Figs. 6.85, 6.86), kidney degeneration and haemorrhagic nephritis (Fig. 6.87).

Fig. 6.74 Mute swan (*Cignus olor*) naturally infected with HPAI of the H5N1 subtype exhibiting torticollis. (Courtesy of Antonio Camarda)

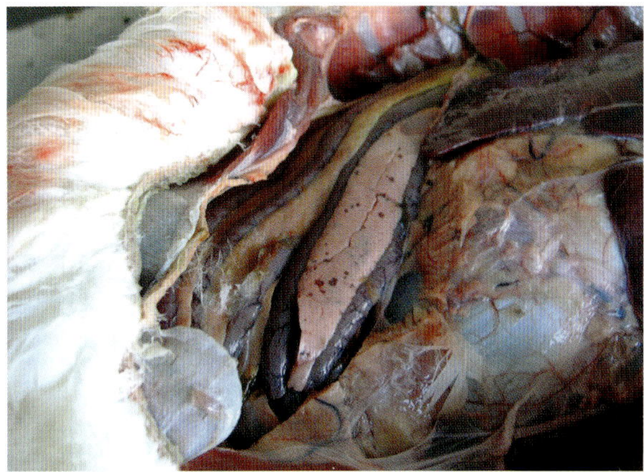

Fig. 6.75 Mute swan (*Cignus olor*) naturally infected with HPAI of the H5N1 subtype, displaying pancreatitis with petecchial haemorrhages. (Courtesy of Daniel Baroux)

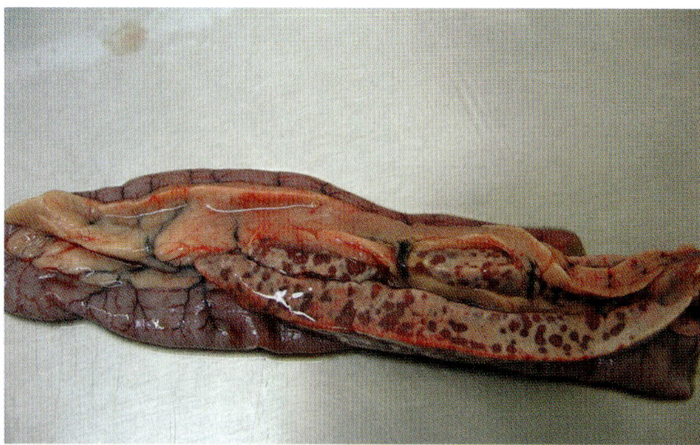

Fig. 6.76 Mute swan (*Cignus olor*) naturally infected with HPAI of the H5N1 subtype, displaying pancreatitis with petecchial haemorrhages. (Courtesy of Daniel Baroux)

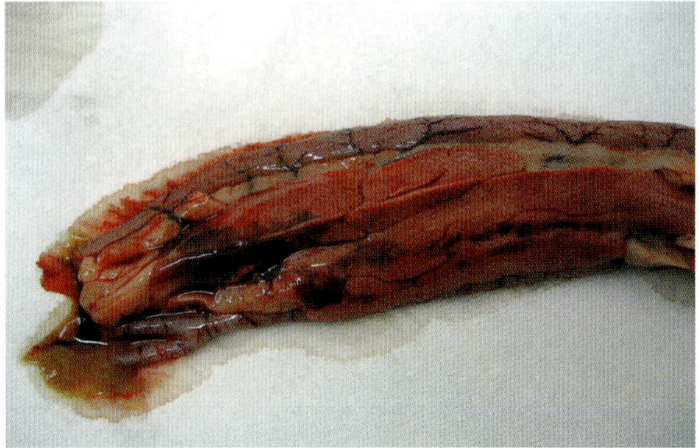

Fig. 6.77 Mute swan (*Cignus olor*) naturally infected with HPAI of the H5N1 subtype, presenting haemorrhagic pancreatitis and duodenitis. (Courtesy of Daniel Baroux)

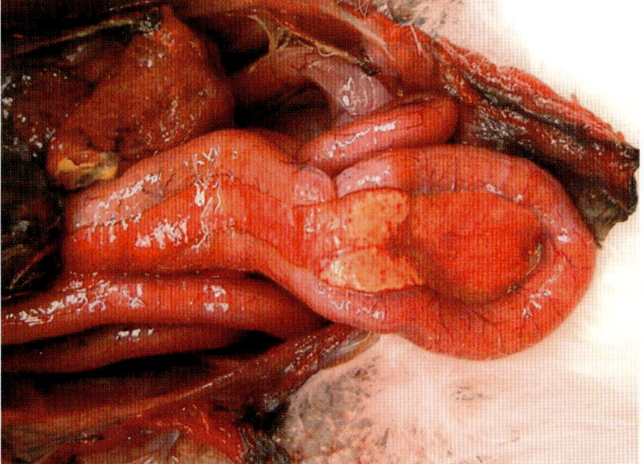

Fig. 6.78 Mute swan (*Cignus olor*) naturally infected with HPAI of the H5N1 subtype, showing necrotic and haemorrhagic pancreatitis. (Courtesy of Caroline Brojer)

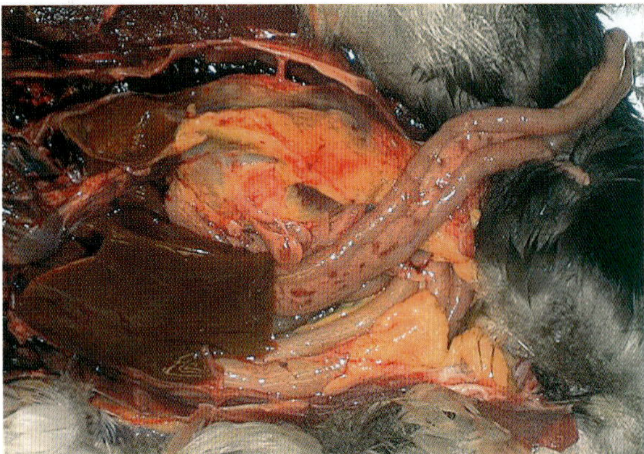

Fig. 6.79 Tufted duck (*Aythya fuligula*) naturally infected with HPAI of the H5N1 subtype, presenting necrotic and haemorrhagic pancreatitis. (Courtesy of Daniel Baroux)

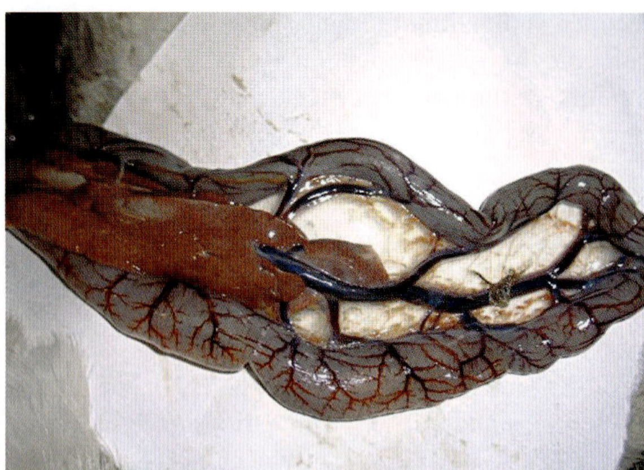

Fig. 6.80 Pelican (*Pelecanus* spp.) naturally infected with HPAI of the H5N1 subtype, displaying duodenitis and pancreatitis. (Courtesy of Victor Irza)

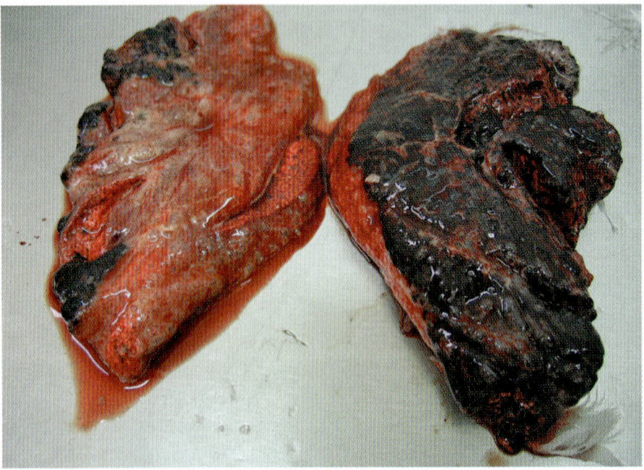

Fig. 6.81 Mute swan (*Cignus olor*) naturally infected with HPAI of the H5N1 subtype, presenting bilateral pneumonia with haemorrhages and oedema. (Courtesy of Daniel Baroux)

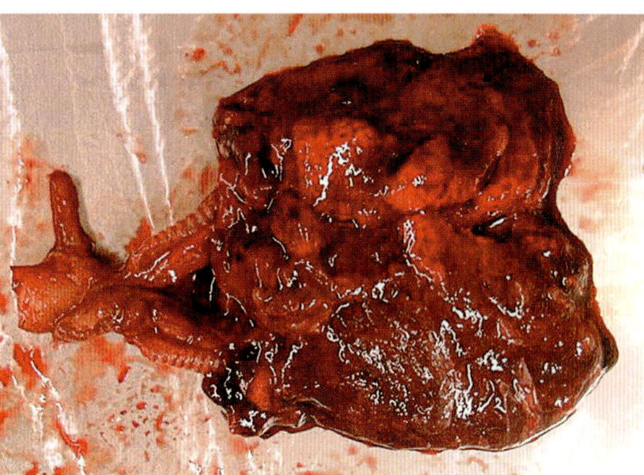

Fig. 6.82 Eagle owl (*Bubo bubo*) naturally infected with HPAI of the H5N1 subtype, displaying bilateral pneumonia with oedema. (Courtesy of Caroline Brojer)

Fig. 6.83 Common pochard (*Aythia ferina*) naturally infected with HPAI of the H5N1 subtype, exhibiting pneumonia, air sacculitis and blood clots in the visceral cavity. (Courtesy of Daniel Baroux)

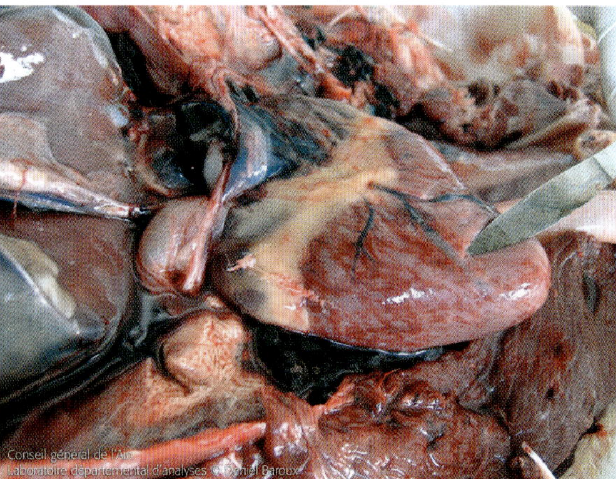

Fig. 6.84 Mute swan (*Cignus olor*) naturally infected with HPAI of the H5N1 subtype, displaying multifocal haemorrhages in the myocardium. (Courtesy of Daniel Baroux)

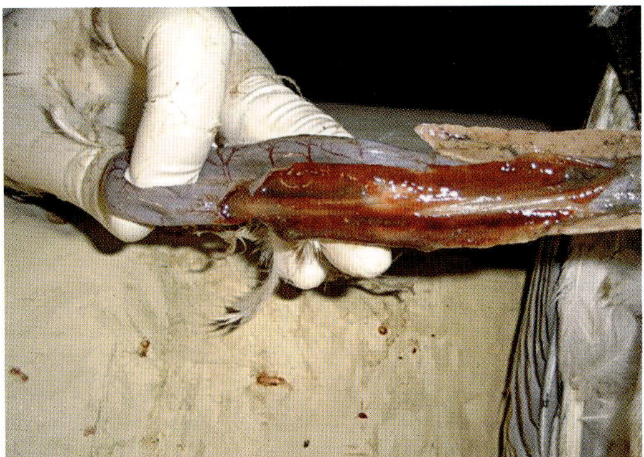

Fig. 6.85 Mute swan (*Cignus olor*) naturally infected with HPAI of the H5N1 subtype, presenting haemorrhagic duodenitis. (Courtesy of Victor Irza)

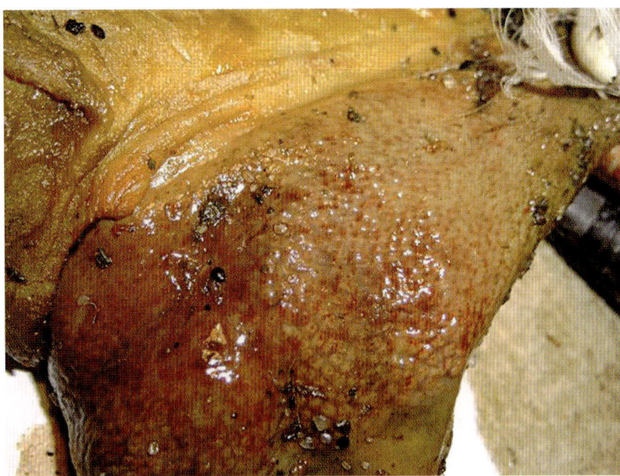

Fig. 6.86 Pelican (*Pelecanus* spp) naturally infected with HPAI of the H5N1 subtype, exhibiting haemorrhages in proventriculus. (Courtesy of Victor Irza)

Fig. 6.87 Mute swan (*Cignus olor*) naturally infected with HPAI of the H5N1 subtype, presenting with hypertrophic, congested and haemorrhagic kidneys. (Courtesy of Daniel Baroux)

6.4 Guidelines for Farm Visits

6.4.1 Protective Measures During the Collection of Samples in an Infected or Suspected Infected Poultry Farm

In order to minimise the risk of human exposure to AI viruses and to avoid contamination of clothing, which can become a means of spreading AI to other premises, it is essential to take the following precautions:

- In the handling of infected or suspected infected poultry and contaminated poultry materials (e.g. body parts, body tissues, blood, feathers, poultry excretions and other potentially infected materials such as bedding) and when infected birds are slaughtered or during cleaning and disinfection operations, care should be taken to prevent or minimise the generation of dust and other aerosols.
- Farming premises containing infected or suspected infected poultry should only be entered by trained staff, with the number limited to only those needed to carry out essential operations.

6.4.1.1 Individual Protection Devices (IPDs)

When poultry rearing areas are entered, special clothing and personal protective gear should be worn; these must be removed before personnel leave the area and then stored in tightly closed containers for professional cleaning/disinfection or disposal, so that the spread of virus is prevented. Such items include:

1. Body-covering workwear (overalls, disposable body suits, disposable underwear)
2. Headgear that completely covers the hair
3. Boots that can be disinfected
4. Watertight, protective gloves that can be disinfected
5. Mouth and nose protection or, if aerosol formation cannot be reliably prevented, a tight-fitting breathing mask (FFP2 or FFP3)
6. Eye protection, for example close-fitting goggles with side protection

A breathing hood (THP3) is preferable to separate breathing and eye protection. After the workwear/protective clothing is removed, a whole-body shower should be taken and the hands then disinfected. Specific legal requirements relating to animal disease control must be observed.

6.4.2 Summary of Information to be Collected from a Suspect Index Case (see also Annex 2: Epidemiological Inquiry Form)

Before the diagnosis of AI is attempted and prior to necropsy procedures, it is essential that a complete history of the outbreak is collected, including:

- Identification of the holding suspected to be the centre of infection and preliminary identification of the productive units and subunits involved in the outbreak; location of the main suspected holding and density of poultry farms in the area; biosecurity measures applied by each unit and subunits involved; characteristics of the different production systems (backyard flock or intensive farm); proximity to wetlands hosting resident or migratory waterfowl.
- Identification of staff involved in managing the unit.
- Anamnestic data, including information on food consumption and the egg production rate of domestic poultry (layers, turkey and broiler breeders infected with HPAI virus may initially lay soft-shelled eggs, but will soon stop laying), duration of the clinical signs and the number of sick or dead birds. Relevant information that may help to elucidate the origin of the disease must include the location and time of appearance of disease in affected birds.

Every susceptible species present in the farm must be clinically investigated, beginning from the most peripheral units to the centre of the outbreak. Particular attention must be paid to whether some or all of the birds have been vaccinated. All this information must be reported in the epidemiological inquiry. An official report must identify the species of all the animals present in each unit, the starting date and the description of the clinical signs and the percentage mortality.

6.4.2.1 Clinical Examination

A careful clinical examination of live affected birds before culling is essential to identify the characteristic signs caused by very virulent viruses, including neurological signs such as: incoordination, tremor, torticollis, abnormal gait and paralysis. In addition, HPAI often results in depression, blindness, severe enteritis with haemorrhagic diarrhoea and severe respiratory problems.

6.4.2.2 External Examination

During physical examination of the birds, it is very important to observe the feathers around the vent to recognise signs of diarrhoea and to inspect unfeathered areas such as legs, joints, comb and wattles. Birds affected by HPAI typically have cyanotic and oedematous combs and wattles, which may have petechial or ecchymotic haemorrhages at their tips. Subcutaneous haemorrhages may also be seen. Specific note should be made of any swelling on the infraorbital sinuses and of cloacal discharges. The latter should be qualified as to nature, colour, consistency and odour.

6.4.2.3 Necropsy Precautions

At least five birds suspected of being infected with HPAI must be subjected to post-mortem examination in diagnostic laboratories by competent authorities and in full compliance with health and safety regulations.

The necropsy table must be dampened with a soap and water solution. The same solution should be used to dampen the feathers of the birds to be necropsied, in order to minimise the formation of dust and aerosols. Resistant rubber gloves, a respiratory mask with an aspiration valve (FFP2 or FFP3) and a protective visor or protective glasses must be worn.

All necessary equipment, i.e. knives, forceps, scissors, plastic containers for samples, swabs etc, should be within reach in the necropsy room.

Bins for the disposal of necropsied birds must be in sufficient number to contain all waste.

6.5 Differential Diagnosis

6.5.1 Summary of the Main Post-mortem Lesions Associated with HPAI and LPAI

Infections in birds can give rise to a wide variety of clinical signs that may vary according to the host, virus strain, host immune status, presence of exacerbating secondary organisms and environmental conditions.

HPAI in poultry may cause a sudden high flock mortality of up to 100%, even without preliminary clinical signs. Birds dying of the peracute form of avian influenza show minimal gross lesions, consisting mainly of dehydration and congestion of the viscera and muscles.

Lesions in the respiratory tract, when present, are a result of severe, acute respiratory distress as manifested by coughing, gasping and expectoration of a bloody exudate. In birds that die after a more prolonged clinical course, petechial and ecchymotic haemorrhages occur throughout the body, particularly in the larynx, trachea, proventriculus, caecal tonsil and epicardial fat, and on the serosal and mucosal surfaces of respiratory and digestive organs. Often there is extensive subcutaneous oedema, particularly around the head and hocks. The carcase may be dehydrated. Yellow or grey necrotic foci may be present in the spleen, liver, kidneys and lungs. The air sacs may contain an exudate. The spleen may be enlarged and haemorrhagic.

Generally speaking, LPAI is asymptomatic in wild birds, while in poultry without exacerbation it is responsible for only mild to severe respiratory signs. In layers, LPAI infection may be responsible for a drop in egg production and a decrease in the hatching rate. Chickens and turkeys may show ruffled feathers, apathy and a drop in food and water consumption. Acute diarrhoea may be present without loss of weight.

Diseases that must be considered in the differential diagnosis of LPAI are those causing a drop in egg production and mild respiratory signs. These are infectious bronchitis, infectious coryza, pneumovirus infection, laryngotracheitis in its mild form and mycoplasmosis.

It must be taken into account that LPAI often coexists with a secondary bacterial or mycoplasma infection. In the former, LPAI may evolve into a chronic form in which there is a loss of weight, swelling of the infraorbital sinuses, severe respiratory signs and to flock mortality of 40–70%, particularly in young turkeys.

Pasteurella multocida and *Escherichia coli*, when associated with LPAI, are often responsible for macroscopic lesions such as fibrinous air sacculitis, pericarditis and abdominal egg retention.

6.5.2 Differential Diagnosis of Highly Pathogenic Avian Influenza

Any circumstance causing a sudden high mortality must be considered in the differential diagnosis of HPAI. Infectious diseases such as velogenic strains of ND virus, acute poisoning or husbandry mismanagement should be taken into account. The latter include extreme alterations of heating, ventilation, or humidity, all of which may lead to very severe consequences when animals are kept in a closed rearing system. Mismanagement of temperature regulation may result in the production of carbon monoxide. Acute poisoning may produce sudden death in a high percentage of animals. Other forms of poisoning may be due to ingestion of pesticides, intoxication caused by an overdose of anti-coccidial drug, or chlorine intoxication due to careless disinfection of the environment.

Sudden high mortality may also be due to botulism, which is an intoxication caused by ingestion of the toxins of *Clostridium botulinum*. Botulism has been observed both in poultry and in waterfowl due to contaminated feed. Signs appear within a few hours to a few days following ingestion of the toxin. In chickens, these signs include weakness and progressive loss of control of the legs, wings and neck. Tremors and paresis progress to paralysis. The most severely affected birds die within a few hours.

The main infectious diseases that need to be differentiated from HPAI are briefly reviewed below:

6.5.2.1 Newcastle Disease

The disease most similar to AI is ND, which is caused by different strains of avian paramyxovirus type 1. ND viral strains are characterised according to the degrees of pathogenicity (e.g. velogenic, mesogenic and lentogenic strains). For a more detailed description of the disease, see the specific chapter dedicated to ND. The differentiation of a velogenic form of ND from HPAI is virtually impossible without laboratory diagnosis.

6.5.2.2 Infectious Laryngotracheitis

Avian infectious laryngotracheitis is a globally distributed poultry disease caused by a herpesvirus that induces severe respiratory distress. Chickens and pheasants are the most frequently affected species, although infections in other birds such as peafowl have also been reported. The incubation period ranges from 5 to 10 days. To date, only one serotype has been reported although strains show significant variability in their pathogenicity. Most strains are markedly virulent. Depending on the virus pathotype, the disease may present as either an acute or subacute form. The acute form spreads rapidly within the flock; death is sudden and flock mortality may be as high as 50%. Characteristic signs are severe dyspnoea with open-mouth breathing, loud gasping and the coughing-up of haemorrhagic secretions. Haemorrhagic secretions can often also be found scattered around the walls of premises. A drop in food and water intake and a reduction in egg production are common.

Post-mortem findings are limited to the upper respiratory tract, consisting of haemorrhagic laryngotracheitis with blood clots and mucoid exudates in the trachea. Cyanosis of the carcase is also often present. Infectious laryngotracheitis is a notifiable disease in most countries; when suspected, animal health authorities must be notified immediately.

6.5.2.3 Infectious Bursal Disease (Gumboro Disease)

Infectious bursal disease (IBD) is caused by very virulent strains of IBD virus (VVIBDV) belonging to the family *Birnaviridae*. IBD is an acute, contagious, viral disease of young chickens. Other forms of IBD may be much milder and the virus may infect other birds without causing any clinical signs. In case of virulent IBD, morbidity and mortality begin 3 days post-infection and then peak, receding over a period of 5–7 days.

Clinical signs are present in chicks > 3 weeks of age and are characterised by diarrhoea and dehydration. Trembling, incoordination and vent pecking usually occur. Chicks also exhibit depression, anorexia and ruffled feathers. The virus severely damages the bursa of Fabricius and lymphoid organs such as the thymus, caecal tonsils and spleen. The infection produces variable degrees of immunosuppression resulting in a high susceptibility to secondary pathogens. If husbandry is poor or the virus strain is particularly virulent then mortality in the flock may exceed 30%.

Gross lesions are the enlarged and haemorrhagic bursa, which tend to atrophy after the acute phase, increased presence of mucus in the intestine, swelling of the kidneys with urate deposits and foci of necrosis in other lymphoid organs. Petechial haemorrhages may be present in the thigh and pectoral muscles and occasionally at the junction of the proventriculus and gizzard.

IBD is a notifiable disease in most countries; when suspected, animal health authorities must be notified immediately.

6.5.2.4 Duck Virus Enteritis (Duck Plague)

Duck virus enteritis (DVE) is an acute herpesvirus infection of ducks that causes haemorrhagic enteritis, in some cases leading to sudden death. Flock mortality can be 100% within 1–5 days. The incubation period is between 3 and 5 days after exposure.

The acute form of DVE is characterised by haemorrhagic and watery diarrhoea. Other clinical signs are dehydration, weakness, lethargy, inappetence and a marked drop in egg production. Infected birds sometimes present with tremors and ataxia. The feathers of the birds are ruffled and blood-stained.

Post-mortem findings are severe enteritis, haemorrhages and crusty plaques on the mucosa of the entire digestive apparatus, from the oesophagus to the intestinal tract and including the caeca and rectum.

6.5.2.5 Acute Fowl Cholera

Avian cholera is a contagious disease resulting from infection by *Pasteurella multocida*. Suspicion of avian cholera must be considered when a large number of dead animals are found in a short period of time. All domestic species are susceptible to the disease, but ducks, turkeys and quail are the most likely to be infected. Acute pasteurellosis can result in death 6–12 h after exposure. Susceptibility to infection and the course of the disease depend upon many factors: sex, age, genetic variation, immune status, concurrent infections, nutritional status and strain virulence.

Some birds appear lethargic while others may show neurological signs, such as convulsions, swimming in circles or erratic flight due to inflammation of the middle ear. Mucous discharge from the mouth, blood-stained droppings from the nose and ruffled feathers are some of the other clinical signs. Mortality from avian cholera in poultry may exceed 50% of the affected flock. Death can occur very rapidly, in which case gross lesions are generally absent. A commonly observed lesion in birds that have died of fowl cholera is a congested, enlarged and haemorrhagic spleen. Frequent findings also include haemorrhages on the surface of the heart, liver and gizzard. In addition, areas of liver tissue with yellow spots, alteration of texture, colour and shape can be found. *Pasteurella multocida* may also be localized to the wattles and comb, resulting a swollen and oedematous appearance.

Avian cholera is a notifiable disease in most countries; when identified, the animal health authorities must be notified.

7 Conventional Diagnosis of Avian Influenza

Calogero Terregino and Ilaria Capua

7.1 Isolation of Avian Influenza Virus

7.1.1 Sample Management and Preparation

Assign an identification number to the sample as soon as it arrives at the laboratory.

Fill out a work sheet indicating the identification number of the sample, kind of sample, species, test to perform and starting date of analysis.

This work sheet should be prepared and maintained by the technician who performs the tests.

7.1.1.1 Swabs

- Under sterile conditions and using a laminar flow cabinet, transfer 2 ml of PBS solution containing antibiotics from the tube in which the swabs have been soaked into another tube.
- In case of highly contaminated swabs add 2 ml of antibiotics to obtain a final dilution of 1:2.
- Centrifuge the sample at 1,000 × g for 10 min to remove coarse particulate matter.
- Use the supernatant as an inoculum for the specific-pathogen-free (SPF) embryonated eggs.
- Keep the sample at 4°C overnight, or store it at –80°C if there are longer delays in processing until eggs can be inoculated.

7.1.1.2 Organs

- Thaw organs at room temperature or at 4°C.
- Working under a laminar flow cabinet and using sterile forceps and scissors, collect an amount of sample corresponding to 1 cm^3. Organs can be pooled according to the apparatus they belong to.
- Homogenise the sample in a mortar, adding sterile quartz sand.
- Add 9 ml PBS with antibiotics.
- Decant the homogenate into a 15-ml tube. Leave the sample at 4°C for 1 night. If the eggs will not be inoculated the next day, store the sample at –80°C until the date of the test.
- Centrifuge the sample at 1,000 × g for 10 min.
- Use the supernatant to infect embryonated eggs.

7.2 Virus Isolation

Virus is isolated following the protocol of the OIE and according to European standards (OIE Manual 2008, EC 94/2005).

7.2.1 Methods

1. Candle 9- to 11-day-old embryonated SPF fowl's eggs to check embryo viability. Mark the shell of the egg with a pen to delimit the air sac. With a manual or electric device, drill a small hole just above the air sac (Fig. 7.1).
2. Record the identification number of the sample, the passage number (1° or 2°), the kind of sample (lung, cloacal swabs, etc.) and the date of inoculation on five eggs.
3. Inoculate 0.1–0.2 ml of clarified supernatant obtained from the tracheal and cloacal swabs or from the organ homogenate into the allantoic cavity of each of the five 9- to 11-day-old embryonated SPF fowl's eggs (Fig. 7.2 a-d).
4. Seal the eggs with glue or wax (Fig. 7.3).
5. Incubate the inoculated eggs at 37°C for 7 days.

I. Capua, D.J. Alexander (eds.) *Avian Influenza and Newcastle Disease*,

6. Candle the inoculated eggs daily to check embryo vitality.
7. The number of dead eggs for each sample should be recorded daily in the lab's working handbook.
8. Test the allantoic fluid of eggs containing dead embryos for haemagglutinating (HA) activity as indicated below (Fig. 7.4).
9. If HA activity is detected, identify the HA agents by means of the haemagglutination inhibition (HI) test as described below.
10. After 7 days, chill the remaining eggs in a refrigerator (4°C) to end the first passage.
11. The following day, open the eggs using sterile techniques under a laminar flow cabinet.

Fig. 7.1 Perforation of specific-pathogen-free (SPF) eggs for virus isolation attempts

Fig. 7.2 a–d (**a**) Inoculation of SPF eggs for virus isolation attempts. (**b**) Needle introduced into the allantoic cavity, above the border of the air cell (*black horizontal line* on the egg shell). (**c**) Embryonated chicken egg: inoculation via the allantoic route. (Courtesy of Amelio Meini). (**d**) Anatomy of an embryonated chicken egg at 9–10 days of incubation: schematic view. (Courtesy of Amelio Meini)

Collect approximately 10 ml of the allantoic fluid.

12. Test the allantoic fluid for the presence of HA activity by the "rapid HA test" as described below.

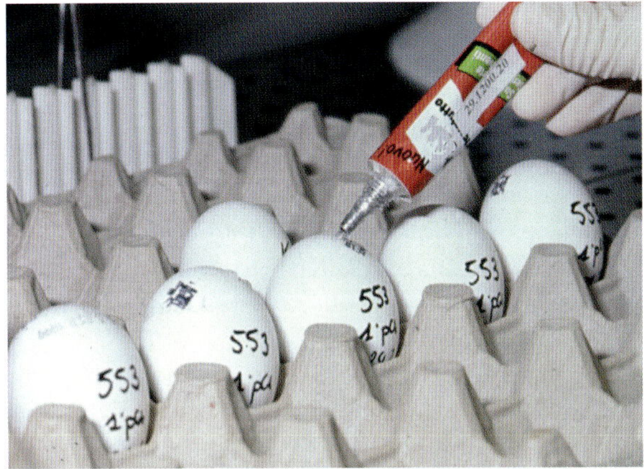

Fig. 7.3 Sealing of SPF eggs for virus isolation attempts

13. If at the end of the first blind passage, no HA activity is detected, use the undiluted allantoic fluid, collected from this passage, to perform a second passage in embryonated eggs as described above.
14. The mortality of inoculated eggs 24-h post-inoculation is generally considered non-specific, although some HPAI viruses, may cause embryo mortality as early as 18 h post-infection.
15. If no HA activity is detected following two passages in eggs, the sample may be considered negative.
16. When HA activity is detected, the presence of bacteria must be excluded by culture.
17. If bacteria are present, the fluids must be passed through a 450-nm membrane filter, with further addition of antibiotics, and inoculated into embryonated eggs as described above.
18. A sample giving a positive result in the rapid HA test must be tested further. These tests in-

Fig. 7.4 a-d Rapid haemagglutination reaction. Incubation time: 5 s (**a**), 10 s (**b**) 20 s (**c**) and 30 s (**d**) clude the titration of HA activity. This procedure enables the confirmation and quantification of HA, which are prerequisites to the identification of the HA agent by means of the HI test.

If a laboratory does not have the capacity to perform the HI test, the HA allantoic fluid should be sent to a National Reference Laboratory or to an International Reference Laboratory to confirm diagnosis.

7.2.2 Haemagglutination Test in Petri Dishes (Rapid HA Test)

This method is based on the reaction between the HA activity of the virus and red blood cells (RBCs). If viral replication has occurred in the embryonated SPF egg, the allantoic fluid will contain virus particles with HA activity. The latter can be visualised by adding a drop of allantoic fluid to a drop of a RBC suspension; the resulting reaction is macroscopically visible.

1. Place in a Petri dish a drop of the allantoic fluid with the same amount of 1% RBC suspension.
2. Allow the two fluids to mix.
3. Wait 30–60 s and observe whether there are any RBC aggregates (Fig. 7.4 a-d).

7.2.3 Characterisation of Avian Influenza Viruses

7.2.3.1. Haemagglutination Test in Microtitre Plates (Micro HA Test)

1. Dispense 0.025 ml PBS into each well of a plastic microtitre plate (V-bottomed wells).
2. Place 0.025 ml of virus suspension (i.e. allantoic fluid) in the first well.
3. Use a multi-channel micropipette to make twofold dilutions (from 1:2 to 1:4096) of virus suspension across the plate. Discard the last 0.025 ml.
4. Dispense 0.025 ml of PBS into each well.
5. Add 0.025 ml of 1% RBC to each well.
6. Mix by tapping gently and place at 4°C or at room temperature (20–24°C).
7. Plates are read after 30 min (at room temperature) or after 40 min (at 4°C) when the RBC controls have settled and assumed a button shape. If HA activity is present, it will appear as a fine layer of RBCs lining the entire bottom of the well. If HA activity is absent, RBCs will settle at the centre of the well in the shape of a button. The samples are read by holding the plate perpendicular to the bench, in other words by holding it vertically, against a white background and observing the presence or absence of tear-shaped streaming of the RBCs (Fig. 7.5). In wells

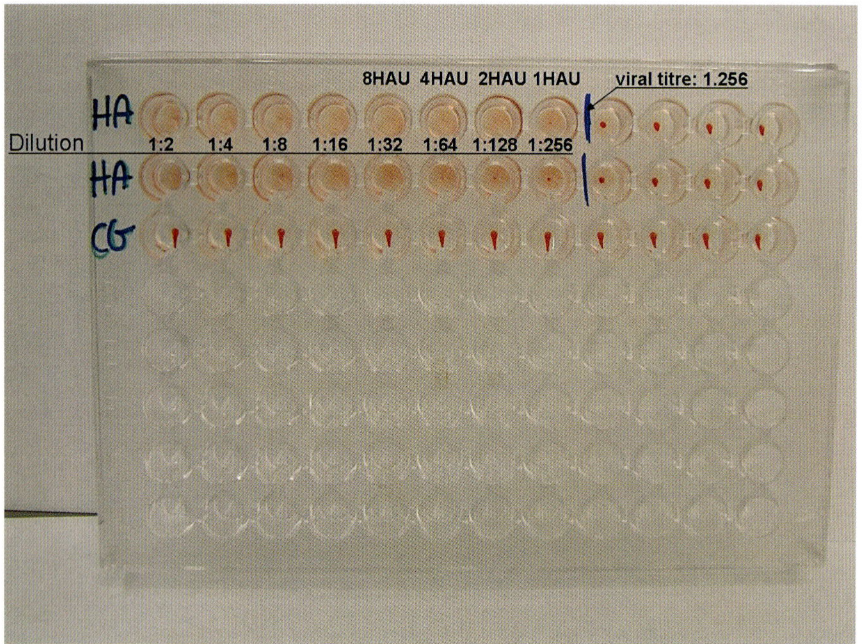

Fig. 7.5 Haemagglutination reaction in microplates

with HA activity, there will be no streaming. In wells without HA activity, the RBCs should flow at the same speed as the RBCs in the control wells in which only PBS and RBCs have been dispensed.

The HA titre of the viral suspension under examination is the highest viral dilution that causes agglutination of the RBCs (no streaming). This dilution is defined as containing one haemagglutinating unit (HAU). To perform the HI test and characterise the virus subtype, a 4-HAU antigen solution is conventionally used, i.e. containing four times that viral concentration. For example: if the HA titre obtained is 1:512 (1 HAU), 4 HAU will be obtained by dividing that titre by 4 (512:4=128). A dilution of 1:128 of the haemagglutinating allantoic fluid will be used to prepare the antigen solution for the HI test. For accurate determination of the HA content, this should be done from a close range of an initial series of dilutions, i.e. 1/3, 1/5, 1/7 etc.

7.2.3.2 Haemagglutination Inhibition Test (HI Test)

This method is based on a reaction between the virus and the specific antiserum. When the antiserum reacts with the virus, it binds to the epitopes that are responsible for haemagglutination; thus, these epitopes will not be available to bind to RBC. If the antiserum is not specific for the virus, haemagglutination will occur, indicating non-identity between the two reagents.

Reagent Preparation

- *Reference antiserum*: Reconstitute one vial of freeze-dried serum with 1 ml of sterile distilled water or according to the manufacturer's instructions. Prepare 16 antisera against all the different known AI subtypes and include one against Newcastle disease (ND) virus. Unused reconstituted antisera may be aliquoted and stored at –20°C.
- *Virus to be characterised*: Prepare the viral suspension with 4 HAU as described above.

Procedure

1. Dispense 0.025 ml PBS into all wells of a plastic microtitre plate with V-bottomed wells, except the first well of the 4-HAU control row, generally H1–H6.
2. Place 0.025 ml of each reference antiserum in the first wells of the first column of the plate (A1–G1) and use the last row (H) to titrate the 4 HAU and for the RBC control (Fig. 7.6).
3. Use a multichannel micropipette to obtain two-fold dilutions of all sera across the plate and discard the last 0.025 ml.
4. Add 0.025 ml of diluted allantoic fluid containing 4 HAU in each well from row A to row G.
5. In the first two wells of the last row (4-HAU virus titration control: H1–H6) of each plate, dispense 0.025 ml of diluted allantoic fluid containing 4 HAU and make two-fold dilutions from the second well to the sixth well (H2-H6). Discard the last 0.025 ml.
6. Add 0.025 ml of PBS in all wells of the virus control and 0.050 ml of PBS in the RBC control wells (H7–H12).
7. Mix by tapping gently and place the plate at +4°C for 40 min or at room temperature for 30 min.
8. Add 0.025 ml 1% RBCs to all wells.
9. Mix by gentle tapping and place at 4°C or at room temperature. Plates are read after 30–40 min, when the RBCs control has settled. This is done by holding the plate in a perpendicular position to the bench, in other words by holding it vertically, against a white background and observing the presence of tear-shaped streaming at the same speed as that occurring in the RBCs control wells.

Results

In the first three wells (H1–H3) of the 4-HAU control, haemagglutination must be observed. In well H4, a partial haemagglutination (half of a tear-shaped drop) and in wells H5 and H6 no haemagglutination should be seen. Wells H1–H6 correspond to 4 HAU, 2 HAU, 1 HAU, 0.5 HAU, 0.25 HAU and 0.125 HAU respectively.

The virus is identified on the basis of the correspondence with the reference antiserum, which inhibits its haemagglutinating activity. In case of identity, the titre of the reference antiserum with the virus under examination should be equal to or ± 1 dilution of its titre with a homologous antigen (Ag) (see Fig. 7.6).

The 4HAU control ensures that the correct amount of Ag has been used in the test.

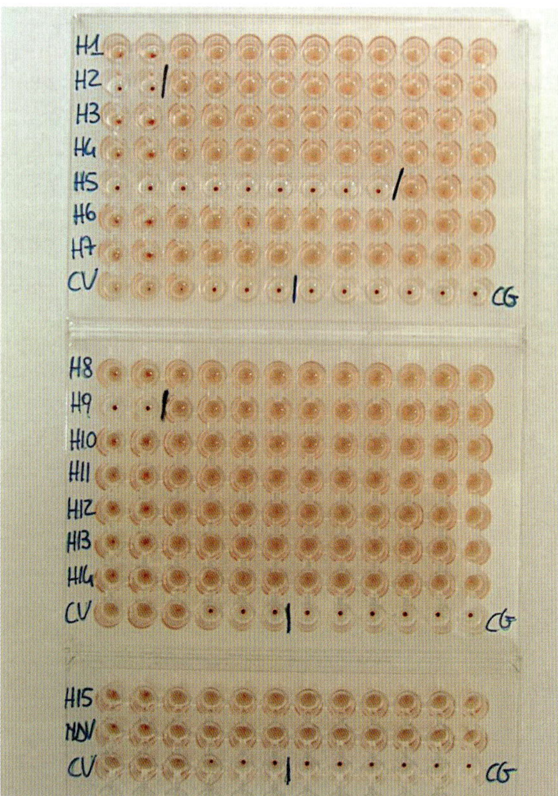

Fig. 7.6 Typing of avian influenza virus by the haemagglutination inhibition test. The strain was identified as H5 subtype. Weak cross-reactions with H2 and H9 antisera were observed. *CV* 4 HA unit control; *CG* red blood cell control

7.2.3.3 Neuraminidase Inhibition Test

The neuraminidase inhibition (NI) test is a laboratory procedure aimed at characterising the neuraminidase subtype of influenza viruses. The assay is based on the reaction between the unknown neuraminidase viral subtype and monovalent antisera able to inhibit the enzymatic activity of neuraminidase. Nine different neuraminidase subtypes (N1–N9) of influenza type A are known and all of them have been associated with isolates obtained from birds.

This complex method is based on inhibition of neuraminidase protein activity with monovalent antisera raised against each of the neuraminidase subtypes. The enzymatic activity of neuraminidase is detected through the release of N-acetyl neuraminic acid from the fetuin substrate. Subsequently, the addition of periodate oxidises the N-acetyl neuraminic acid into β-formyl pyruvic acid, while the addition of thiobarbituric acid (TBA) develops a chromophore that can be extracted with acid butanol. The results may be read colorimetrically using a spectrophotometer, with the optical reading proportional to the neuraminidase activity in the original preparation. The enzymatic activity will be inhibited only by the specific antiserum against the neuraminidase subtype of the virus that is being characterised.

In practical terms, if the antiserum is not specific for the neuraminidase, enzymatic activity is retained, resulting in the production of a pink-coloured solution. If the specific antiserum binds to the neuraminidase protein, then enzyme activity is inhibited and the chromophore remains white (colourless).

This test should be performed only after the HA subtype of the virus has been characterised.

Safety

The NI test releases toxic compounds during the immersion of tubes in boiling water. Thus, this phase should be carried out under a fume hood and operators should wear a protective mask.

Equipment

- *100- to 300-μl pipettes*
- *Laminar-flow cabinet*
- *37 ± 2°C incubator*
- *56–100°C water bath*
- *Spectrophotometer (549 nm) (optional)*
- *Plastic test tubes 12×75 mm*
- *Glass test tubes 16×100 mm*
- *Syringes*
- *Rubber plugs*
- *Rack for pipettes*
- *10-ml pipettes*

Working Reagents for Neuraminidase Inhibition Assay

- *Antisera against each of the nine different neuraminidase subtypes. The antisera should be stored at –20°C*
- *Standard fetuin substrate*
- *0.1M PBS (pH 5.9)*
- *0.025M Sodium periodate in 0.125 NH_2 SO_4*
- *2% Sodium arsenite in 0.5N HCl ($NaAsO_2$)*
- *0.1M TBA (pH 9.0)*
- *5% 10N HCl butanol acid (optional)*
- Instructions for the preparation of these solutions are provided in Annex 4.

Procedure

1. Source antisera for each different subtype of neuraminidase (9 in total). The haemagglutinin of the sera should be different from that of the virus being tested. Example: if the virus under examination is H7, then antisera H1N1, H1N2, H5N3, H8N4, etc. should be chosen. Do not select antisera raised against H7 viruses.
2. Dilute the antisera 1:5 in PBS (pH 5.9)
3. Dilute the virus 1:15 in PBS (pH 5.9) if the HA titre is > 1:64. Dilute the virus 1:13 if the HA titre is ≤ 1: 64.
4. Set up the antisera and the sample as shown in the scheme below:

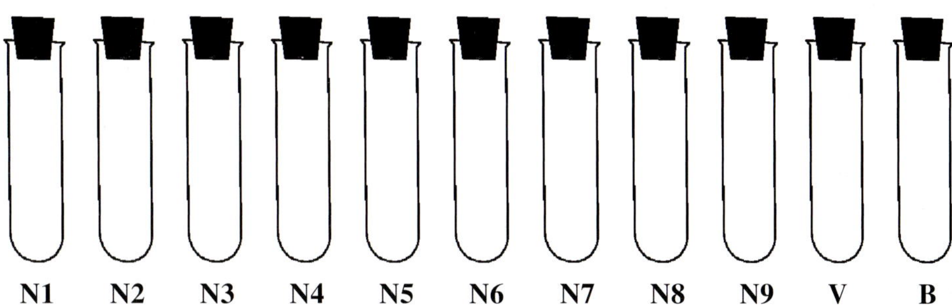

V Virus control, *N* neuraminidase subtype, *B* blank (reaction control)

5. Dispense 100 µl of the diluted antiserum into each glass test tube (serum anti-N1 into tube N1, serum anti-N2 into tube N2, etc.).
6. Add 100 µl of diluted virus to all tubes except tube B (blank).
7. Add 100 µl of PBS (pH 5.9) into tube V (virus control) and 200 µl into tube B (reaction control).
8. Seal each tube with its plug and incubate the closed tubes at room temperature for 30 min.
9. Add 300 µl of standard fetuin to each tube. Carefully shake the tubes for 15 s to mix the reagents.
10. Seal the tube with its plug and incubate at 37°C for 16–20 h.
11. Add 200 µl of sodium periodate to each tube. Carefully shake the tubes for 15 s in order to allow the reagents to mix.
12. Seal each tube with its plug and incubate at 37°C for 30 min.
13. Add 200 µl of sodium arsenite to each tube. Carefully shake the tubes to mix the reagents. A brown colour will develop and then disappear. It is possible at this stage to stop the assay by storing the tubes at +4°C.
14. Add 2 ml of TBA to each tube. Carefully shake the tubes for 20 s to mix the reagents.
15. Remove the plug and partially immerse the tubes in boiling (100°C) water for 7.5 min.

Results

Tubes containing a pink coloured solution = no inhibition has occurred. The antiserum used is not specific for the neuraminidase.

Tubes containing colourless or a white-coloured solution compared to virus control tube = total inhibition. The antiserum used is specific to the neuraminidase.

For example, if a white/colourless solution has been obtained using antiserum against N2 then the neuraminidase is type 2 (Fig. 7.7).

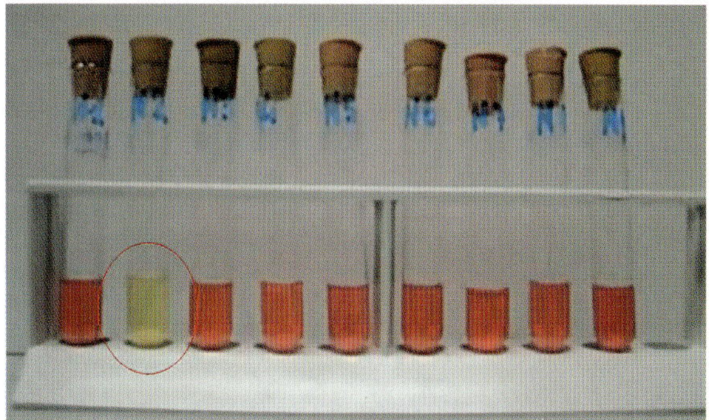

Fig. 7.7 Typing of avian influenza virus by the neuraminidase inhibition test. The strain was identified as N2 subtype (*red circle*)

7.2.3.4 In vivo Pathogenicity Test

The virulence for chickens of influenza A viruses isolated from birds must be assessed using the intravenous pathogenicity index (IVPI) test.

Reagents and Equipment

- *Allantoic fluid containing virus to be assessed*
- *Sterile isotonic saline*
- *10 SPF or specific-antibody-negative (SAN) 6-week-old chickens*
- *1-ml syringes*
- *Test tubes, cotton wool*
- *Negative-pressure isolators*
- *Feed, litter and drinking water*
- *Disposable bags*

Procedure

1. Dilute fresh infective allantoic fluid with a HA titre > 1:16 1:10 in sterile isotonic saline. The fluid should be from the lowest passage available, preferably from the initial isolation and without prior selection.
2. Inject 0.1 ml of the diluted virus intravenously into each of ten 6-week-old SPF or SAN chickens.
3. The birds are examined at 24-h intervals for 10 days. At each observation, each bird is assigned a score of 0 if normal, 1 if sick, 2 if severely sick and 3 if dead.
   The judgement of sick vs severely sick is a subjective clinical assessment. Normally, 'sick' birds show one of the following signs and 'severely sick' more than one of the following signs:
   - *Respiratory involvement*
   - *Depression*
   - *Diarrhoea*
   - *Cyanosis of the exposed skin or wattles*
   - *Oedema of the face and/or head*
   - *Nervous signs*

Dead birds must be scored as 3 at each of the remaining daily observations after death. On welfare grounds, when birds are too sick to eat or drink, they must be killed humanely and scored as dead at the next observation since they are expected to die within 24 hours without intervention. This approach is acceptable to accreditation authorities.

The IVPI is the mean score per bird per observation over the 10-day period. An index of 3.00 means that all birds died within 24 h while an index of 0.00 means that no bird showed any clinical signs during the 10-day observation period.

A simple method for recording results and calculating indices is shown in the following example:

Clinical signs	Days after the inoculation Number of chickens with specific signs										Total score
	1	2	3	4	5	6	7	8	9	10	
Normal	10	0	0	0	0	0	0	0	0	0	10 × 0 = 0
Sick	0	6	0	0	0	0	0	0	0	0	6 × 1 = 6
Severely sick	0	2	4	2	0	0	0	0	0	0	8 × 2 = 16
Dead	0	2	6	8	10	10	10	10	10	10	76 × 3 = 228
											TOTAL = 250/100 IVPI = 2.50

10 birds observed for 10 days = 100 observations
Index = mean score per bird per observation = 250/100 = 2.50.
Any influenza A virus, regardless of subtype, yielding a value > 1.2 in an IVPI test is considered to be a highly pathogenic avian influenza (HPAI) virus.
Data from Commission Decision of 4 August 2006 approving a Diagnostic manual for avian influenza as provided for in Council Directive 2005/94/EC (2006). Official Journal of the European Union L 237/1-27.

7.3 Serology

7.3.1 Introduction

Domestic and wild birds infected with AI viruses through natural exposure or vaccinated against AI viruses develop antibodies that can be detected by means of serological tests. The antibody response to AI viruses in birds is directed against the immunogenic proteins of the virus. Natural infection with AI viruses elicits the production of both type-specific (against type A matrix (M) and nucleoprotein Ag) and subtype-specific (against the HA and N proteins) antibodies. It is therefore essential to implement a diagnostic protocol while bearing in mind the diagnostic significance of the result. Tests that are able to detect antibodies against type A Ag are group-specific, i.e. detection of these antibodies indicates that the bird has been infected (or vaccinated) with an influenza A virus of any subtype. These tests are the agar gel immunodiffusion (AGID) assay and the enzyme-linked immunosorbent assay (ELISA), both of which are directed against the NP and/or M proteins.

Subtype-specific antibodies are detected by means of the HI test and indicate that infection (or vaccination) has occurred with a virus of a specific H subtype. The HI test gives no indication of the N subtype of the virus that has elicited the immune response, as the antibodies measured by this test are only those targeting the haemagglutinin.

Generally speaking, serological diagnosis of AI is performed in two steps, unless the subtype circulating in a given area is already known. The first step aims at the detection of antibodies to any AI virus and is implemented by applying the AGID assay or the ELISA. The second step, performed on samples determined to be positive in step 1, identifies the viral subtype causing infection.

7.3.1.1 Step 1: Detection of Group Antigen (Type A) Antibodies

These tests are able to detect antibodies to the group Ag of influenza A viruses. The antigen-antibody reaction is against the NP or M proteins of AI viruses. These Ags are present in all influenza A viruses regardless of the H or N subtype. Serological positivity to these tests indicates that the birds have encountered an influenza A virus, but no information on the AI subtype that has caused seroconversion can be deduced.

Note: Serological positivity to type A in waterfowl (wild and domestic) is a very common finding.

Agar Gel Immunodiffusion (AGID) Assay

This is a simple and reliable test in chicken and turkey sera. It is very specific but is of limited sensitivity; thus, it must be used as a diagnostic tool on a flock basis. It can be performed in any laboratory with basic equipment. It is completely unreliable in waterfowl, as these birds do not produce precipitating antibodies. It has not been validated in other avian species.

Enzyme-Linked Immunosorbent Assay (ELISA)

This is a test that requires more advanced laboratory equipment, including a spectrophotometer. It is highly sensitive but lacks specificity. In the case of indirect ELISA tests, care must be taken to ensure that the secondary antibody in the test (anti-species) is directed against the species under examination. Competitive ELISA tests have the advantage that serum of any species can be examined. The instructions of the assay's manufacturer should be followed and assumptions on test reactivity for species other than those mentioned in the kit's specifications should be avoided.

7.3.1.2 Step 2: Detection of Subtype-Specific (H Subtype) Antibodies

This test is used in birds known to be infected with AI virus – either following a positive serological test against the group (type A Ag) or as a result of clinical history. The test identifies the haemagglutinin subtype of the virus causing the seropositivity. The HI test is used for this purpose. Low-level cross-reactivity with other H subtypes may be observed due to homology with the neuraminidase Ag. This cross-reactivity is generally not higher than 1:16 (2^4) and disappears with another Ag containing a different neuraminidase subtype. For example: A serum sample is positive to H9N2 at a titre of 1:256 (2^8). If tested with H5N2 Ag, a positive inhibition result is observed at 1:8 (2^3). When tested with an H5N9 antigen the sample will be negative.

7.3.2 Agar Gel Immunodiffusion (AGID) Assay

This test is widely and routinely used to detect the presence of antibodies against influenza A virus in the serum of birds. Since it is very specific but is of limited sensitivity, it is used as a diagnostic tool on a flock basis. Antibodies against influenza A viruses are detected by the lines formed following precipitation of the immune complex established between antibody in the test sera and the reference Ag.

7.3.2.1 Preparation of Agar Dishes

1. Dissolve 8 g NaCl in 100 ml distilled water in a volumetric flask.
2. Add 1.25 g Noble agar and mix gently.
3. Dissolve the agar by immersing the flask in a boiling-water bath until the agar is completely dissolved.
4. Transfer 15 ml of the agar solution to each 90 mm Petri dish.
5. Leave the dishes uncovered and allow the agar to cool at room temperature.
6. Label the batch of agar dishes with the production date on each lid, then seal the dishes in an air-tight plastic bag. The agar dishes can be stored up-side-down (to avoid the condensation of water droplets on the lids) for up to 15 days at +4°C.

7.3.2.2 Test Procedure

1. Record the identification number of the samples on the dishes.
2. Punch wells with the agar punch as shown in the Figures 7.8, 7.9. Remove the agar plugs with a steel tip or a Pasteur pipette attached to a vacuum pump.
3. Place 30 µl Ag into the central well.
4. Add 30 µl positive antiserum (S+) into two wells that are directly opposite from each other (Figs. 7.8, 7.9).
5. Place 30 µl of the serum under examination (SE) in each of the remaining wells. The layout of the reagents ensures that each SE is adjacent to a well containing a positive serum S+ and to one containing Ag.
6. Incubate the dishes in a humid chamber at room temperature for 48 h.

7.3.2.3 Interpretation of the Results

1. Read the plates 48 h post-incubation using a diffuse light source, with illumination from below the plate.
2. The test is valid when a precipitation band is seen between the wells containing the positive serum and the central well containing the Ag (Figs. 7.8, 7.9).
3. The sample is positive when a precipitation band is observed between the well containing the test serum and the central well containing the Ag. The band must be continuous (i.e. show identity) with the precipitation band formed between the positive serum and the Ag.
4. The sample is negative when no precipitation band is seen between the well containing the SE and the central well containing the Ag.

Fig. 7.8 Agar gel precipitation assay. *Ag* Reference antigen, *S+* positive serum, *SE* serum under examination

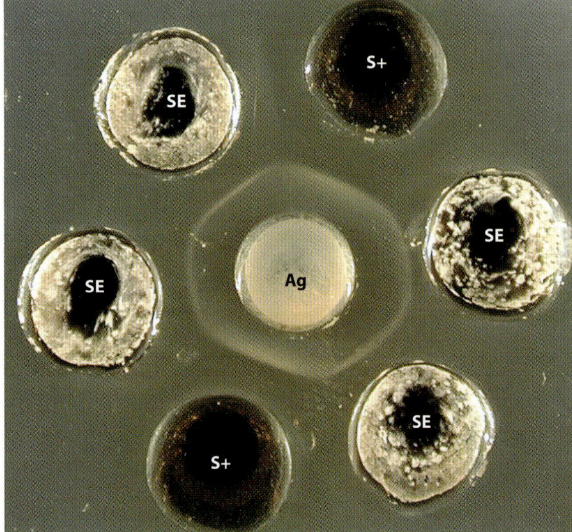

Fig. 7.9 Agar gel precipitation assay: *Ag* Antigen, *S+* positive serum, *SE* serum under examination

7.3.3 *Enzyme-Linked Immunosorbent Assay (ELISA) To Detect Antibodies Against Avian Influenza Virus Type A*

An ELISA is a useful and sensitive test that can be applied for serological screening. A variety of different kits are now available on the market, allowing detection of the presence of specific antibodies against the NP and the M of proteins AI viruses. The instructions of the assay kit's manufacturer must be followed carefully when performing the test.

7.3.4 *Detection of Specific Antibody Subtypes by the Haemagglutination Inhibition (HI) Test*

This method is based on a reaction between a HA virus and an antiserum containing specific antibodies to that virus. When the HA virus is incubated with its specific antibodies, the natural HA activity of the virus is inhibited. This is visualised by adding to the well a suspension of RBCs, which will sediment in the shape of a button.

Since the HI test is both qualitative and a quantitative, a known amount of antigen, in most cases 4 HA units, must be used.

7.3.4.1 Preparation of Red Blood Cell Suspensions for HA and HI Tests

A 1% RBC suspensions is used in the HA and HI assays of chicken sera, while a 10% RBC suspension is used in the pre-treatment of sera originating from species other than chickens. Nonspecific reactions, due to the presence of nonspecific HA in the sera of species other than chickens, may occur during the HI assay. To avoid this, non-chicken sera are pre-treated with chicken RBCs to remove non-specific HA.

Procedure

1. Collect 5 ml of blood from at least three SPF chickens with a syringe containing sufficient Alsever's solution to yield a ratio of 1:1.
2. Pool the contents of the syringes and centrifuge the blood suspension at 1,000 g × 10 minutes. Discard Alsever's solution or supernatant.
3. Wash the RBCs twice in PBS solution, centrifuging at 1,000 g × 10 minutes after each washing.
4. The supernatant is removed with a pipette, leaving the packed RBCs which are then used to prepare a RBC suspension of appropriate concentration for a given test. The suspension can be stored at +4°C for 7 days.

10% RBC suspension: Prepare 9 ml of 0.05% bovine albumin PBS solution and add 1 ml of packed RBCs.

1% RBC suspension: Prepare 99 ml of 0.05% bovine albumin PBS solution and add 1 ml of packed RBCs. The correct percentage should be checked by one of the following methods.

7.3.4.2 Measurement of the RBC Concentration

Spectrophotometric Method

This system measures the RBC concentration indirectly, by quantifying the haemoglobin content of the solution.

1. Set up two cuvettes each with 3.6 ml of distilled water and 0.4 ml of the suspension to be read in the spectrophotometer at 545 nm wavelength.
2. Refer to the PBS 0.05% albumin solution as the blank.
3. A 1% RBC suspension should result in an optical density (O.D.) of 0.250 nm.

Microhematocrit Tube Method

The haematocrit may be measured manually by centrifugation.

1. Fill a standard capillary microhaematocrit tube with blood and seal the tube at the bottom.
2. Centrifuge the tube at 10,000 rpm for 5 min. The RBCs have the greatest mass and are forced to the bottom of the tube.
3. Measure the height of the red cell column as a percent of the total blood column. The higher the column of red cells, the higher the hematocrit and the percentage.

Cell Counting Chamber

In this procedure, either the Thoma, Burker, or Malassez chamber can be used. In each of these chambers, 75 million to 80 million RBCs/1 ml correspond to a concentration of 1%.

7.3.4.3 Haemagglutination Inhibition Test for Chicken Sera

Reagents

- *PBS*
- *PBS and albumin (PBS/albumin 0.05%)*
- *Freeze-dried reference antigen diluted with PBS to obtain 4 UHA per 0.025 ml*
- *Chicken RBCs suspension (1%)*
- *Negative control chicken serum*
- *Positive control chicken serum*

Note: Reconstituted reference sera must be store at –20°C and reconstituted reference antigens at –80°C.

Procedure

1. Dispense 0.025 ml PBS into all the wells of a microtitre plate, with the exception of the H1 well.
2. Dispense 0.025 ml of serum into the first wells of the microtitre plate (column 1). Add 0.025 ml of the positive control serum (with known HI titre) to the F1 well and 0.025ml of negative control serum to the G1 well. Control sera should be ideally included in each plate or batch of 10 plates. The 4-HAU control and RBCs control should be included in each plate.
3. Using a multi-channel micropipette, make two-fold dilutions of the sera (A1–A12) across the plate. Discard the last 0.025 ml.
4. Add 0.025 ml of antigen suspension containing 4 HAU across the plate, with the exception of row H.
5. Add 0.025 ml of antigen suspension containing 4 HAU to the first two wells of row H, then make two-fold dilutions from H2 to H6 (discard the last 0.025 ml) in order to obtain 4, 2, 1, 0.5, 0.25, 0.125 HAU. Wells H1–H6 contain the 4-HAU control titration.
6. Add 0.025 ml PBS + albumin 0.05% to all wells of the H row.
7. Mix by gentle tapping. Place the plate at +4°C for 40 min or at room temperature for 30 min.
8. Add 0.025 ml of 1% RBCs suspension to all wells.
9. Mix by gentle tapping. Incubate at +4°C for 40 min or at room temperature for 30 min.
10. Read the plates after 30–40 min, when the RBC control has settled. This is done by holding the plate in a vertical position and recording the occurrence or absence of tear-shaped streaming of the RBCs at the same speed as in control wells containing RBCs (0.025 ml) and PBS (0.05 ml) only (Figs. 7.10, 7.11). The HI titre is the highest dilution of serum causing complete haemagglutination inhibition.

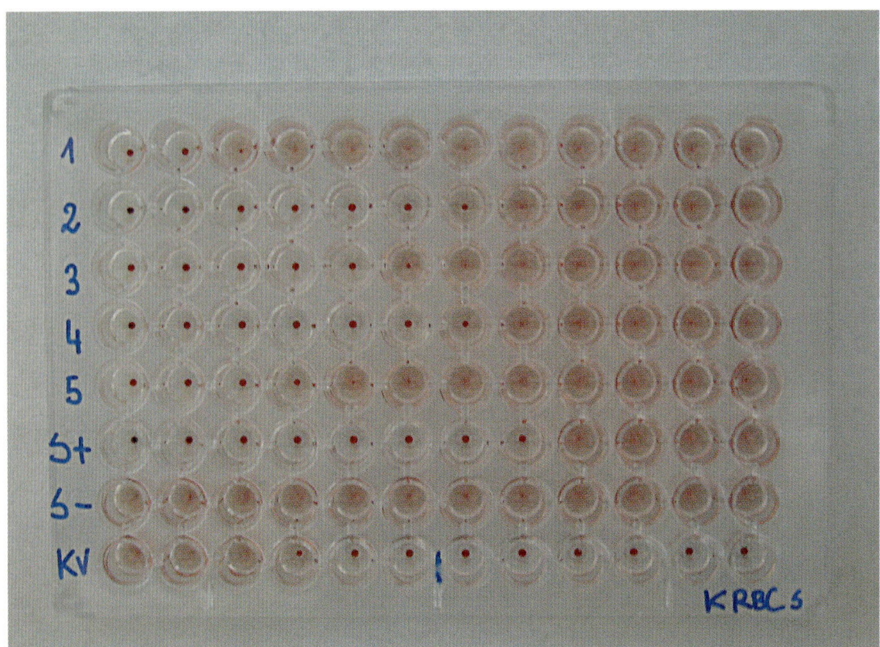

Fig. 7.10 Haemagglutination inhibition assay. *A1–A5* Test sera, *S-* negative serum, *S+* positive serum; the last row is the 4-HA unit control (*KV*) and the last six wells contain the RBCs control (*KRBCS*)

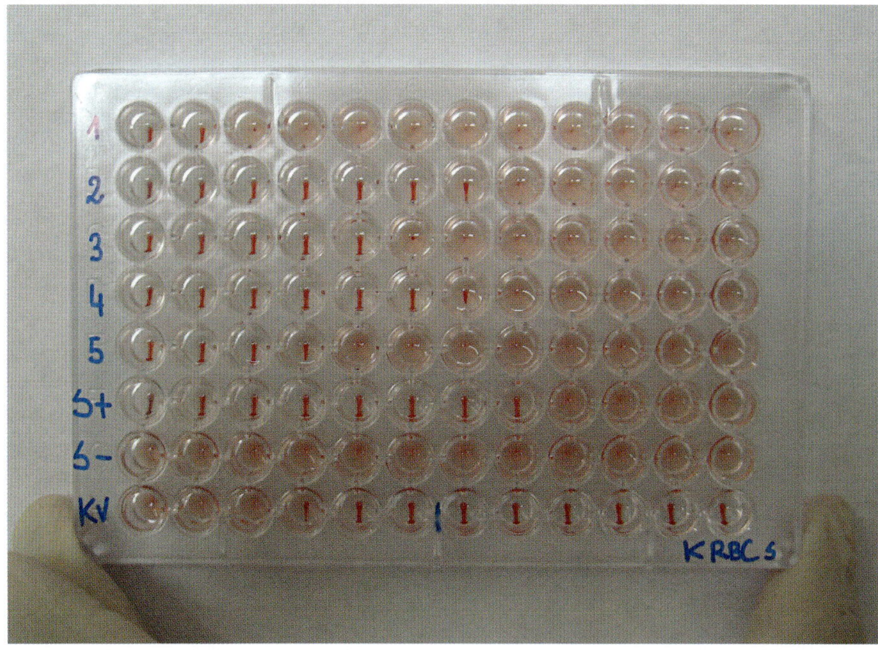

Fig. 7.11 Haemagglutination inhibition assay. *A1–A5* Test sera, *S-* negative serum, *S+* positive serum; the last row is the 4-HA unit control (*KV*) and the last six wells contain the RBCs control (*KRBCS*). Same plate as Fig. 7.10, tilted

Interpretation of the Results

1. The test is valid if:
   - The *negative control serum* has a titre of less than 2^3.
   - The *positive control serum* has a titre that coincides with its declared titre or is within one dilution of its declared titre.
   - Complete haemagglutination is observed in the first three wells (H1–H3) of the 4-HAU control row (containing respectively 4, 2, 1 HAU)
2. The sample is considered negative if the HI titre is < 1:8. This means that the serum does not contain specific antibodies to that subtype of AI virus.
3. The sample is considered positive if the HI titre is $\geq$ 1:16. This means that the serum contains specific antibodies to that subtype of AI virus.

7.3.4.4 Haemagglutination Inhibition Test for Species Other than Chickens

Non-specific reactions frequently occur when sera samples coming from birds other than chickens are used. Sera must be pre-treated as described below.

Pre-treatment of Sera

1. Dispense 0,050 ml PBS into wells in the first column of the microplate (wells A1–E1).
2. Leave the second row (A2–E2) empty.
3. Dispense 0.025 ml PBS into all other wells of the microtitre plate.
4. Add 0.050 ml of test sera to the first wells of the microplate (column 1).
5. Add 0.050 ml of a 10% RBCs suspension to the first wells (column 1).
6. Incubate the plate for 30-40 min at room temperature, and wait for the 10% RBCs suspension to settle.
7. Transfer 0.025 ml of the supernatant from the wells of the first column to the wells of the second column.
8. Transfer an additional 0.025 ml of the supernatant from the wells of the first column to the wells of the third column. Make two-fold serial dilutions of the sera from the third column to the last column (12). Discard the last 0.025 ml.
9. The sera are now ready to be treated as chicken sera. Column 1 should be excluded from the test.

It is also recommended to inactivate the non-specific haemagglutinating agents in the serum of game birds (pheasant, partridge, etc.) and of quails, ostriches and guinea fowl by heat treatment in a water bath at 56°C for 30 min prior to testing.

Molecular Diagnosis of Avian Influenza

8

Giovanni Cattoli and Isabella Monne

8.1 Introduction and Basic Terminology for Molecular Diagnostic Tests

The use of diagnostic methods based on molecular technology has improved substantially over the last decade; consequently, laboratory tests for the identification and characterisation of avian influenza (AI) viruses have become available. Such tests may be used to detect the AI viral genome directly from clinical specimens as well as to generate data on the molecular characteristics of an isolate or of viral RNA present in a sample collected from an infected animal. In addition, the current definitions of highly pathogenic AI (HPAI) (EC 2006; OIE 2004) are based on the results of conventional and molecular techniques, thus, the latter are among the official methods used to identify virulence factors and to confirm the presence of HPAI viruses in laboratory specimens.

8.1.1 *Retrotranscription-polymerase chain reaction (RT-PCR)*

The vast majority of the methods for the molecular detection of AI viruses are based on the retrotranscription (RT) of the viral RNA into a DNA copy (cDNA). The cDNA is then amplified by the polymerase chain reaction (PCR). The entire process is therefore known as RT-PCR.

Conventional RT-PCR protocols, consisting of a RT-PCR amplification step followed by gel electrophoresis of the amplified products, may be applied directly on clinical specimens. Depending on the sets of primers used, the virus type (type A) or subtype (H5 and H7) may be identified. In addition, the amplified cDNA can be sequenced directly, allowing a timely characterisation and pathotyping of the virus detected. The recent introduction of real-time RT-PCR (rRT-PCR) techniques for the detection of AI RNA has led to a more rapid, sensitive and sometimes more specific method to detect the viral genome in clinical samples.

8.2 Two-Step and One-Step RT-PCR

Reverse transcriptase enzymes, necessary for the retrotranscription of RNA into cDNA, are generally obtained or derived from retroviruses. RT-PCR can take place in either a two-step or a one-step reaction.

In two-step RT-PCR, the RNA molecule is first transcribed into cDNA using random oligomers, oligo-dT primers or one target-specific primer. The reaction occurs in a separate tube and under separate reaction conditions. An aliquot of the RT reaction is subsequently added to the PCR for cDNA amplification. The random oligomers or oligo-dT primers allow the total RNA present in the sample to be transcribed, such that several PCRs can be carried out for different targets while using the same cDNA aliquot.

In one-step RT-PCR, RT and PCR amplification take place in the same tube. This is possible due to specialised chemistries and cycling conditions. In this case, only target-specific primers are used, both for the RT and for the PCR. This procedure is faster than the two-step procedure, reduces the number of handling steps and minimises the risk of sample contamination. The advantages of each of these procedures are summarised in the table below (Table 8.1).

Table 8.1 The advantages of two-step and one-step retrotranscription (RT) polymerase chain reaction (PCR)

Two-step RT-PCR	One-step RT-PCR
Multiple PCRs from one RT reaction	Easy to handle
Flexibility in the choice of RT primers	Fast protocols
Long-term storage of cDNA	Good reproducibility Low risk of contamination

I. Capua, D.J. Alexander (eds.) *Avian Influenza and Newcastle Disease*,

8.3 Extraction of RNA from Clinical and Laboratory Samples

Influenza viral genome is negative single-stranded RNA, which can be extracted from a variety of animal specimens (blood, organs, tissues, faeces, swabs, etc.) of infected animals or laboratory culture systems. RNA is an unstable molecule and RNase (enzyme that degrades RNA) is virtually ubiquitous (see Table 8.2). It is important to handle RNA with disposable gloves and to use disposables such as pipette tips and reaction tubes that are certified RNase-free. The use of DEPC (diethylpyrocarbonate, see below) or RNAse-free water is also strongly recommended, although distilled water is generally RNase-free. Alternatively, the use of ready-made RNase-free reagents is recommended. For long-term storage of specimens and extracted RNA, freezing at –80° or in liquid nitrogen will preserve the samples from degradation.

In modern diagnostic laboratories, RNA is extracted using commercially available kits. The major advantages of these commercial kits are:

- The application of a standardised procedure
- High sample throughput and the possibility to automate the extraction procedure
- More time-efficient and less laborious than classical manual procedures
- Use of smaller volumes
- Purity of reagents

The main limitation is the relatively higher cost of the kits.

Traditional phenol-guanidine-based manual methods may also be used to extract viral RNA. This technique can be more efficient in extracting RNA from samples known to yield low amounts of RNA, such as fibrous tissues, and guanidine salts are efficient RNase inhibitors. Moreover, manual systems are cheaper than ready-to-use extraction kits, although they are more labour-intensive and have an increased risk of contamination, which affects downstream applications (such as real-time PCR). Labour-intensive methods are clearly less suitable for processing large numbers of samples.

RNA extraction is a delicate procedure, the success of which influences subsequent steps in molecular testing. Errors at this stage will result in inconsistent or unreliable results (false-negatives and false-positives).

Table 8.2 Main sources of RNase contamination

Origin	Notes
Endogenous RNases	All tissues samples contain endogenous RNases. They should be stored immediately after collection in liquid nitrogen or in special preservative media. Alternatively, they should be immediately placed in chaotropic salt solutions (i.e. lysis buffer included in many commercial kits).
Body surfaces	Skin, bacterial microflora and body fluids such as perspiration are rich in RNase activity. Always use gloves during laboratory activities.
Laboratory surfaces	Environmental contamination by cells from human skin, bacteria and fungi or fungal spores results in the presence of RNase activity on exposed laboratory benchtops, glass- and plastic ware. Treatable material can be cleaned with bleach or commercially available specific products.
Tips and tubes	Autoclaved tips and tubes can still be a source of RNase contamination, since these enzymes are very stable and can re-acquire partial activity after cooling to room temperature. Always use tips and tubes tested and certified as RNase-free.
Waters and buffers	Water and buffers commonly used to prepare laboratory solutions can be a source of RNase. Treat the solution with 0.05–0.1% DEPC or use RNase-free certified reagents. DEPC is a suspected carcinogen. Always wear gloves and handle DEPC under an appropriate fume hood.
RNA storage	Trace amounts of RNase in the eluted RNA samples can cause degradation during storage, even at low temperatures. The best method to preserve isolated RNA for long-term storage (i.e. more than a year at –80°C) is to perform a salt-alcohol precipitation and store the RNA as a precipitate in this solution. Alternatively, commercial solutions that protect RNA during storage are available.

DEPC, diethulpyrocarbonate

8.3.1 DEPC treatment for RNase inactivation

Note: DEPC is a suspected carcinogen. Always wear gloves and handle under a fume cabinet.

Diethylpyrocarbonate reacts with histidine residues of proteins and will inactivate RNases.

1. Add DEPC to solutions at a concentration of 0.05–0.1% (e.g. add 0.5–1 ml of DEPC to 1 l of solution).
2. Stir or shake into solution, incubate for several hours.
3. Autoclave for at least 45 min to inactivate the remaining DEPC.

Note: Compounds containing primary amine groups, such as Tris, will react with DEPC. Tris should be added to the solution only after DEPC treatment is complete.

Note: High concentrations of DEPC or of its by-products can inhibit *in vitro* translation reactions or alter the RNA.

8.3.1 Precipitation Protocols for Partial Purification and Concentration of RNA

8.3.1.1 Ethanol Precipitation (IZSVe Protocol)

1. Add 0.1 volume of 3M NaOAc (pH 5.2) to the nucleic acid solution to be precipitated.
2. Add 2.5 volumes of cold 95% ethanol.
3. Place at –20°C for at least 30 min.
4. Centrifuge at 13,000 rpm at 4°C for 30 min, discard the supernatant.
5. Add 70% ethanol (corresponding to about 4 volumes of the original sample) and centrifuge again at 13,000 rpm for 10 min; discard the supernatant.
6. Add 50 µl of 100% ethanol (corresponding to about 1 volume of the original sample).
7. RNA can then be recovered by centrifugation at 13,000 rpm for 10 min. After precipitation, avoid complete drying of the RNA pellet because it can make the RNA difficult to resuspend. RNA should be resuspended in distilled water or RNase-free TE buffer.

For RNA concentrations < 10 ng/µl, carrier nucleic acid or DNase-treated glycogen should be added to facilitate precipitation and maximise recovery.

8.3.1.2 Lithium Chloride (LiCl) Precipitation (Ambion's Protocol)

Using this protocol, only RNA will be precipitated, while carbohydrates, proteins and DNA will remain in solution. LiCl is frequently used to remove inhibitors of translation that co-purify with RNA prepared by other methods.

1. Add an equal volume of 4 M LiCl, 20 mM Tris-HCl (pH 7.4), 10 mM EDTA to obtain a final concentration of 2 M LiCl.
2. Let the RNA precipitate at –20°C. The time of incubation at –20°C depends on the RNA concentration. As a general rule, it is safe to allow the RNA to precipitate for several hours to overnight.
3. Collect the RNA by centrifugation.
4. Rinse the RNA pellet with 70% ethanol to eliminate LiCl traces (optional). For optimal results, the RNA concentration should be ≥200 µg/ml.

8.4 The Polymerase Chain Reaction

The PCR provides an extremely sensitive means of amplifying a specific sequence of DNA. The principle of PCR is simple: a specific fragment of DNA is repeatedly synthesised using the enzyme DNA polymerase I derived from the bacterium *Thermus aquaticus* (Taq). This organism lives in hot springs and many of its enzymes, including the polymerase, are resistant to thermal denaturation. The thermostability of the Taq polymerase is an essential feature of PCR methodology.

Any region of any DNA molecule (including the cDNA derived from RT reactions) can be chosen, as long as the sequences at the borders of the region of interest are known. This technique has rapidly become one of the most widely used techniques in molecular biology. It is a rapid, versatile, inexpensive and simple means of producing relatively large numbers of copies of DNA molecules (by several-million-fold or more) from minute quantities of source DNA material, even when the source DNA is of relatively poor quality. The development of PCR resulted in a dramatic increase of applications in molecular biology as an increasing number of PCR-based methods were published. For example, PCR can be used for diagnostic tests, DNA fingerprinting, DNA sequencing, screening for genetic disorders, site-specific mutation of DNA, and cloning or subcloning of cDNAs.

A wide variety of samples can be analysed for nucleic acids. PCR protocols use DNA, rather than RNA, as a target because of the stability of the DNA molecule and the ease with which DNA can be isolated. However, it is also possible to start from RNA, providing that this molecule is first transcribed into cDNA. The essential criteria for a sample processed by PCR are that it should contain at least a few copies of an intact RNA/DNA strand encompassing the region to be amplified and that any impurities are sufficiently diluted so as not to inhibit the polymerisation step of PCR.

8.4.1 Components of PCR

Reagents

- *Primers (forward and reverse)*
- *dNTPs*
- *Taq polymerase buffer*
- *Taq polymerase*
- *Magnesium chloride ($MgCl_2$)*
- *Distilled water (ddH_2O)*
- *Primers*: A primer is a short segment of nucleotides that are complementary to the flanking sequences of the DNA segment (template) to be amplified in the PCR. For a PCR, two primers (forward and reverse) are required, each complementary to one strand (indicated as 3'-5' and 5'-3') of the DNA double helix. Primers are annealed to the denatured DNA template to provide an initiation site for the elongation of the new DNA molecule by Taq polymerase. Primer design is extremely important for effective amplification. The primers for the reaction must be very specific for the template to be amplified. Cross-reactivity with non-target DNA sequences results in non-specific amplification of DNA and, at the end, in false-positive results. Also, the primers must not be capable of annealing to themselves or to each other, as this will result in the very efficient amplification of short nonsense DNA molecules and inefficient amplification of the specific target DNA.
- *dNTPs*: The four nucleotide bases, the building blocks of every piece of DNA, are represented by the letters A, C, G, and T, which stand for their chemical names: adenine, cytosine, guanine, and thymine. The A on one strand always pairs with the T on the other, whereas C always pairs with G. The two strands are said to be complementary to each other. The dNTPs are incorporated in the newly synthesised strand by Taq polymerase according to the template strand.
- *Taq polymerase buffer*: This reagent is necessary to create the optimal chemical conditions for the activity of the Taq polymerase.
- *Taq polymerase*: The polymerase recognises the primer-template duplex and begins adding nucleotides to the primer, eventually making a complementary copy of the template. If the template contains an A nucleotide, the enzyme adds on a T nucleotide to the primer. If the template contains a G, it adds a C to the new chain, and so on to the end of the DNA strand.
- *Magnesium chloride ($MgCl_2$)*: This reagent is the catalyst of the enzymatic reaction and is thus crucial for Taq polymerase activity.
- *Distilled water (ddH_2O)*: Reagents are diluted to the correct concentration and the reaction volume is completed with ddH_2O, which should be sterile, free of salts and ions and, for RT-PCR, RNase-free. The latter is essential to avoid degradation of the RNA molecule (see paragraph 8.3).

8.4.2 The Cycling Reaction

There are three major steps in a PCR, which are repeated 30 to 40 times (cycles) (Fig. 8.1). The reaction is done in an automated thermal cycler, which can heat and cool the tubes containing the reaction mixture in a very short time. Most of the common PCR protocols follow the scheme indicated below.

- Denaturation at 94–95°C from 30 s to 1 min. During the denaturation, the double-stranded DNA melts open to single-stranded DNA. All enzymatic reactions (for example, the extension from a previous cycle) are stopped.
- Annealing at X°C from 30 s to 1 min. The primers in the solution are continuously forming and breaking ionic bonds with the single-stranded template. The more-stable bonds (primers that fit exactly) last longer. The polymerase will thus bind those short pieces of double-stranded DNA to begin copying the template. The temperature at which this happens is the annealing temperature. The annealing temperature depends on the primers sequence, but it commonly ranges between 35 and 60°C.
- Extension at 72°C from 30 s to X min (the time depends on the length of the amplified fragment, usually 1 min for 1 kb). This is the ideal working temperature for the polymerase. The bases (complementary to the template) are coupled to the primer on the 3' side (the polymerase adds dNTPs in the 5' to 3' direction).

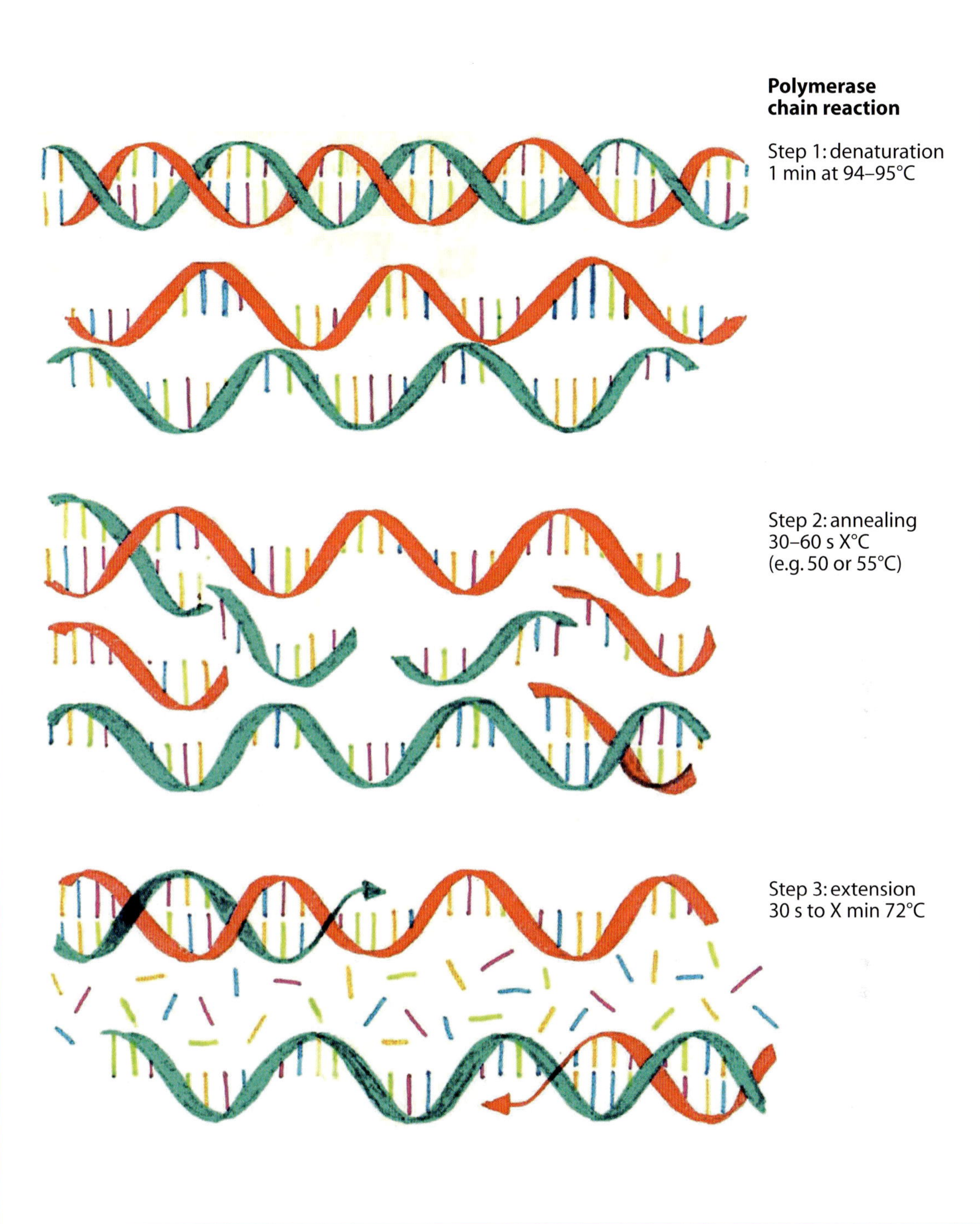

Fig. 8.1 Example of a standard polymerase chain reaction (PCR) cycle. The entire process of amplification by PCR takes 2–4 h, depending on the number of cycles and on the time required for the different steps. In theory, during PCR, a single copy of the target sequence is amplified exponentially. In each cycle, the amount of amplified fragment doubles, such that the correlation factor/ratio between the number of cycles and the amount of amplified fragment is 2^n, where n = number of cycles. Starting from a single target fragment, after one cycle, the number of amplified fragment will be $2^1 = 2$, after two cycles $2^2 = 4$, after three cycles $2^3 = 8$ and so on. After 36 cycles, the number of amplified fragment will be 68 billion copies (see Fig. 8.2). (Courtesy of Amelio Meini)

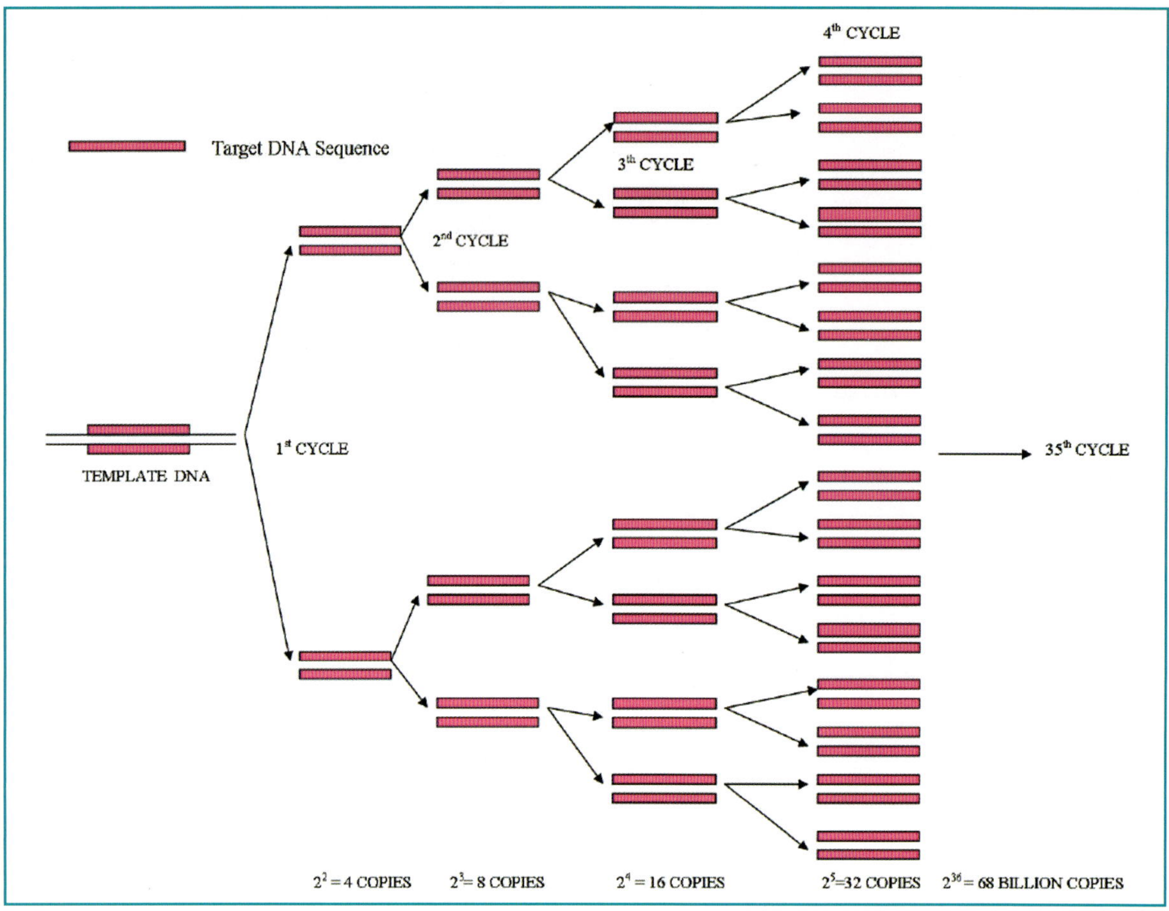

Fig. 8.2 Exponential amplification in PCR

See pages 103–108 for the conventional RT-PCR protocols used in the detection of AI viruses.

8.4.3 Limitations

Like any other laboratory method, conventional PCR has some limitations. Sequence-specific primers are required, meaning that the DNA sequence of the region flanking the amplified DNA must be known. The results of the assay and their specificity can only be determined by gel electrophoresis, which is a relatively time-consuming method involving chemical hazards. In addition, the reaction is limited by the size of the DNAs to be amplified (i.e. the distance between the primers). The most efficient amplification is in the 300- to 1000-bp range, although products up to 4 kb have been amplified.

One of the most important factors to consider during PCR diagnostics is contamination. If the sample that is being tested is even slightly contaminated by target DNA of exogenous origin, this DNA will amplify as well, resulting in false-positive detection. Post-amplification sample handling can also be considered critical with regard to environmental and sample contamination. The huge amount of amplified target DNA molecules can be easily spread to the laboratory environment and to other samples during manipulation of the tubes for electrophoretic detection.

To reduce the impact of some of these limitations of conventional PCR, a new PCR methodology has been developed, called real-time PCR (rPCR if from DNA; rRT-PCR if from RNA).

8.5 Real-Time PCR

In a standard PCR, the amount of product generated should, theoretically, double with each additional cycle. If that were truly the case, then the amount of product present at the end would be a simple function of the amount of template present at the beginning. However, this does not happen; instead, as the reaction progresses, the primers and nucleotides are consumed and eventually become limiting, so the efficiency per cycle gradually drops off. In rPCR, there is a quantitative relationship between the amount of template initially present and the amount of product formed, as shown when a reporter is used to measure the amount of product formed during each individual cycle. The reporter can be linked to a probe (a short oligonucleotide that is complementary to a target sequence located between the primers), or it can be a fluorescent molecule that is able to bind double-stranded DNA (e.g. SybrGreen). Different types of probes are commercially available: hydrolysis probes (TaqMan), hybridisation probes (FRET), molecular beacons, as well as PNA (peptide nucleic acid) and MGB (minor-groove-binding) probes. Despite the difference in their chemistries and modes of action, the basic principle is to measure the fluorescence signal (and then the concentration of the amplified fragment) at the end of each PCR extension step and plot the data. The resulting sigmoid curve will be flat at the baseline level for the first 10–20 cycles (exponential phase) until the amount of template accumulates to the point at which the fluorescence signal becomes detectable. The curve rises linearly for several cycles (linear phase) and finally begins to "roll over", then becoming almost flat (plateau phase, Fig. 8.3). Only in the linear range a direct relationship between the amount of template present in the reaction and the intensity of the fluorescence signal can be assumed. The measurements can be made while the PCR is in progress, hence they are derived in "real time" rather than as an end point (Fig. 8.3). Quantification of DNA, cDNA or RNA targets is easily achieved by determination of the cycle (Ct or Cp, see below for definitions) at which the PCR product is first detected by the machine (i.e. when the product-associated fluorescent signal is intense enough to be automatically detected).

8.5.1 Real-time PCR terms

Background fluorescence: non-specific fluorescence in the reaction, generally more evident during the initial cycles of the PCR. This background signal is commonly used by the machine and related software to determine the baseline fluorescence value.

Threshold: border between the background fluorescence and the specific fluorescence in the reaction. It is usually determined by the machine and its software but also can be set manually.

The **CT (threshold cycle) or Cp (crossing point)** is the cycle number at which the fluorescence signal generated in the reaction crosses the threshold. It is inversely related to the logarithm of the initial copy number of the target sequence.

See pages 109–112 at the end of this chapter for rPCR protocols used to detect AI virus.

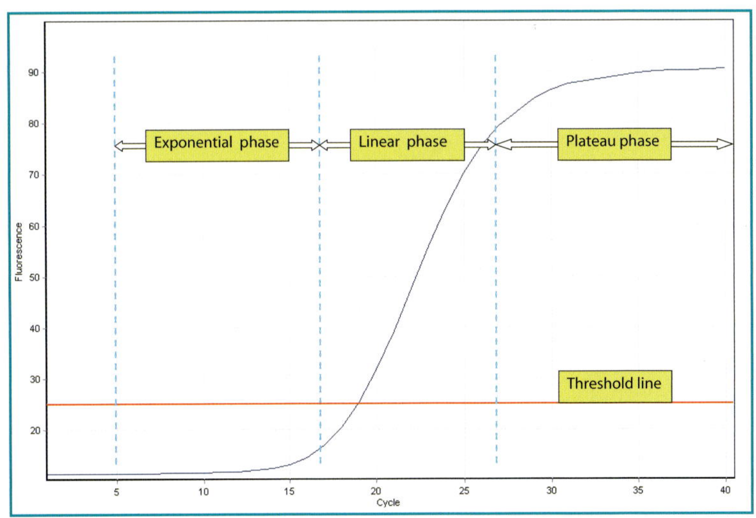

Fig. 8.3 Schematic view of an amplification plot in rPCR

8.6 Organisation of the Laboratory for Molecular Diagnostics

Due to the high sensitivity of the method and the high sample workflow of a PCR diagnostic laboratory, the risk of sample cross-contamination must be taken into account.

For the entire PCR process, the minimal requirement is three separate rooms, organised according to the example below.

First Room: Extraction of Nucleic Acid (DNA/RNA)

Equipment

- Flow cabinet BL2
- Vortex
- Centrifuge
- Micropipettes
- Thermal bath
- Freezer –20°C
- Freezer –80°C
- +4°C Refrigerator

If nucleic acids are to be extracted using organic solvents, such as phenol and chloroform, a fume hood is required.

Second Room: Preparation of Reagents and Reaction Mixes without Template (DNA or RNA)

Equipment

- Flow cabinet
- Centrifuge
- Vortex
- Freezer –20°C (storage of the PCR reagents, primers and probes)

Samples and DNA/RNA templates must not enter this room. Personnel entering this room should wear dedicated lab coats and gloves in order to avoid reagent contamination.

Third Room: Addition of Template Nucleic Acid to the Reaction Mix, Running PCR and rPCR, Gel Electrophoresis Visualisation of the PCR Product

Equipment

- PCR cabinet (to add RNA or DNA)
- Fume hood (for gel electrophoresis chemicals)
- Freezer –20°C
- Refrigerator +4°C
- Thermal cycler and/or
- Real-time PCR platform
- Centrifuge
- Vortex
- Gel electrophoresis unit

8.7 Samples to be Submitted for Molecular Detection of AI and Sample Preparation for PCR

The molecular protocols described in this chapter have been tested on the following sample preparations.

Organs and Tissues (Brain, Trachea, Lungs, Intestine)

- Using sterile scissors or surgical blades, cut small blocks (2–5 × 2–5 mm) of tissues.
- Animal tissues are disrupted with a sterile pestle and mortar. Fibrous tissues (e.g. lung, trachea and intestine) may require the addition of sterile quartz powder or sand to better disrupt the cells. The addition of 300–500 µl of phosphate-buffered saline will facilitate disruption and homogenisation and allow the preparation of sample aliquots for tests other than PCR (e.g. virus isolation).
- Homogenisation is carried out simply using a syringe and needle. Alternatively, commercially available, automatic homogenisers can be used.
- Clarify the suspension obtained by centrifugation at 1.000 g.

Cloacal Swabs, Tracheal Swabs

- Immerse swabs in PBS (maximum 1 ml) and extract the RNA from this suspension. The samples (5 tracheal swabs/pool or 5 cloacal swabs/pool) can be pooled. Vortex briefly.
- Add the appropriate amount of this suspension to the lysis buffer of the RNA extraction kit, according to the manufacturer's instructions.

Faeces

Add 1 volume of faeces to 4 volumes of sterile PBS. Clarify the suspension obtained by centrifugation at 1000 g.
Add the required amount of the supernatant to the lysis buffer of the RNA extraction kit according to the manufacturer's instructions.

Allantoic Fluid

Add the required amount of allantoic fluid to the lysis buffer of the RNA extraction kit according to the manufacturer's instructions.

8.7.1 RNA Extraction

There are several methods for the manual or automated extraction of RNA. Many commercial kits are available, some developed and optimised for the extraction of the nucleic acids from specific samples, such as tissues, blood and stool. Some commercially available extraction kits are used below.

- These kits (or equivalent kit) can be used on different matrices with satisfactory results.
- NucleoSpin RNA II (Macherey-Nage, Germany; catalogue number FC140955N).
- RNeasyMiniKit (Qiagen, Germany).
- High pure RNA isolation kit (Roche Applied Science, Germany): not recommended for faeces.
- MagMax (Ambion/Applied Biosystems): for swabs or other liquid matrices.

The extraction protocol recommended by the manufacturer should be employed.

8.8 Preparation of the PCR Mix

- Before the reaction mix is prepared, it is necessary to calculate the correct volumes of reagents to be used, as shown in the protocols included herein (see examples in Table 8.3). To ensure homogeneous distribution of the reagents in each of the aliquots, it is advisable to mix all the reagents in a single reaction tube (master mix), then aliquot the necessary volume into the PCR tubes in which the reaction for each single sample will take place.
- All of the stock solutions of the reagents for PCR are stored at –20°C. Aliquots of the reagent stock solutions (particularly primers and probes) should be prepared to avoid possible contamination of the stocks and to minimise damage in case of contamination. The enzymes (reverse transcriptase, Taq) should be taken out of the freezer (–20°C) at the last possible moment, to maintain their shelf life.
- The correct volume is added to the appropriate sterile vial.

Procedure

1. Thaw all of the reagents, except for Taq polymerase and/or RT; if possible, keep them on ice.
2. Add the ddH_2O.
3. Add the Taq polymerase buffer.
4. Add the $MgCl_2$.
5. Add the dNTPs.
6. Add the forward primer.
7. Add the reverse primer.
8. Add the Taq polymerase (and/or RT).
9. Vortex the solution.
10. Distribute the single-reaction mixture into separate sterile PCR tubes (as many as the number of samples to be tested).
11. Add nucleic acid template in the flow cabinet in a separate room.
12. Load the vials on the thermal block of the thermal cycler (or real-time PCR platform) and start the reaction.

cont.

Table 8.3 Example for calculating reagent volumes in a PCR setup for ten samples

Reagent (concentration of stock solution)[a]	Final concentration[b]	1× Reaction mix (μl)[c]	Total volume (μl)[e]
Sterile, RNase-free water	/	3.2[d]	35.2
PCR buffer (5×)	1×	5	55
$MgCl_2$ (25 mM)	2.5 mM	2.5	27.5
dNTPs mix (10 mM)	1 mM	2.5	27.5
DTT (100 mM)	10 mM	2.5	27.5
Primer M25F (5 pmol/μl)	0.3 μM	1.5	16.5
Primer M124R (5 pmol/μl)	0.3 μM	1.5	16.5
RNase inhibitor (20 U/μl)	10 U	0.5	5.5
Reverse transcriptase (50 U/μl)	15 U	0.3	3.3
Ampli Taq GOLD (5 U/μl)	2.5 U	0.5	5.5
Final reaction volume: RNA			
Vortex the mix for a few seconds.			
Aliquot 20 μl of the master mix in 0.2-ml PCR tubes.		20	220
(master mix volume for 10 samples +1)[e]			
Add RNA[a].		5	
Final reaction volume[b]		25	/

[a]Generally provided with the product or prepared in the laboratory.
[b]Generally described in the kit's manual or in the PCR protocol (e.g. scientific journal).
[c]For a rapid and easy calculation of the specific reagent's volume necessary to perform one reaction, the following formula can be applied:

$C_i \, x \, V_i = C_f \, x \, V_f$

where C_i is the initial concentration of the reagent (i.e. the concentration of the stock solution provided by the company or prepared in the laboratory). Known.
V_i is the initial volume of the reagent (i.e. the volume necessary to perform one PCR). Unknown.
C_f is the final concentration of the reagent necessary in the PCR. Known.
V_f is the final reaction volume. Known.
For the example above, the $MgCl_2$ volume necessary to perform one PCR (V_i) is:
$25 \text{ mM} \times V_i = 2.5 \text{ mM} \times 25 \text{ μl}$

$$V_i = \frac{2.5 \text{ mM} \times 25 \text{ μl}}{25 \text{ mM}} = 2.5 \text{ μl of } MgCl_2 \text{ (25 mM) to obtain a final concentration of 2.5 mM in 25 μl}$$

[d]Obtained by subtracting the sum of the volumes of the individual reagents from the final volume of the reagents. [μl of water = 20 μl – Σ μl of the different reagents (PCR buffer; MgCl2; dNTP; DTT; primers; RNase inhibitor; RT, Taq].
[e]μl/1 reaction × total number of reactions + 1 (an extra volume is usually calculated to compensate for pipetting errors or imprecise volume dispensations).

8.9 Detection and Analysis of the Reaction Product in Conventional PCR (End-Point PCR)

The final PCR products (also known as an *amplicon*) should be cDNA fragments of defined length. The simplest way to check for the presence of these fragments is by gel electrophoresis (see below). A sample taken from the reaction product (generally a few microlitres) is loaded, along with a ladder comprising the appropriate molecular-mass markers, onto either an agarose gel containing 0.8-4.0% ethidium bromide or a polyacrylamide gel. DNA bands on the agarose gel are visualised by ultraviolet trans-illumination (Fig. 8.4). The bands in an acrylamide gel can be visualised by silver staining. A comparison of the product bands with bands of the known molecular-mass markers allows identification of the product fragments on the basis of their molecular masses.

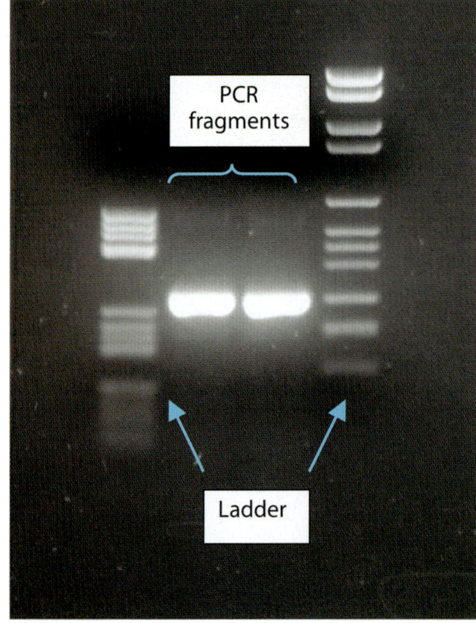

Fig. 8.4 Agarose gel detection of PCR amplicons

8.9.1 Gel Electrophoresis

Electrophoresis is the separation of DNA fragments of different sizes through an agarose or acrylamide matrix placed inside an electrophoresis chamber. The chamber has two opposing electrodes, set on opposite ends and attached to an AC power supply. An appropriate electrophoresis buffer fills the chamber and conducts electricity between these two electrodes. When current is applied, DNA, which is negatively charged, migrates towards the positive electrode. The gel matrix (agarose or acrylamide) acts as a sieve, such that smaller-sized fragments migrate faster than larger ones, thus separating fragments by size. The main difference between agarose and acrylamide gel is the size of the mesh formed by the matrix. The speed of electrophoresis is dependent on the size of the gel and the amount of voltage applied to the gel box by the power supply. The higher the voltage, the faster the migration of the fragments. Each gel box has a maximum optimal voltage range; exceeding this range results in smearing of the DNA bands. Lower voltages generally give cleaner band separation. After electrophoresis, the DNA can be viewed by staining with ethidium bromide or SyBrGreen for agarose gels, and with silver nitrate for acrylamide gels. Ethidium bromide is the stain most commonly used by diagnostic laboratories because of its sensitivity to DNA and the speed of staining. Drawbacks include the cost involved in its visualisation (it requires a UV light source) and its suspected carcinogenicity. However, with appropriate personal protective equipment (PPE, lab-coats, disposable gloves), the low concentrations required for staining minimise any risk.

Silver-stained acrylamide gels are more sensitive and thus allow the detection of smaller amounts of amplicons. The smaller size of the acrylamide gel matrix leads to better separation of the amplicons, particularly very small ones (50–200 bp). These features will contribute to increase the sensitivity and the specificity of the PCR protocol (Fig. 8.5a, b) and ob-

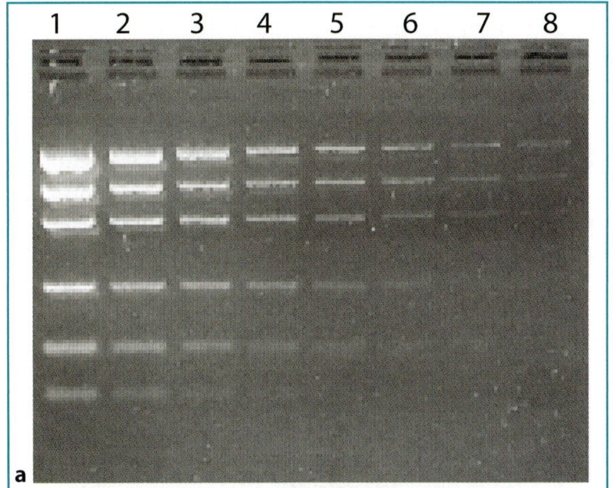

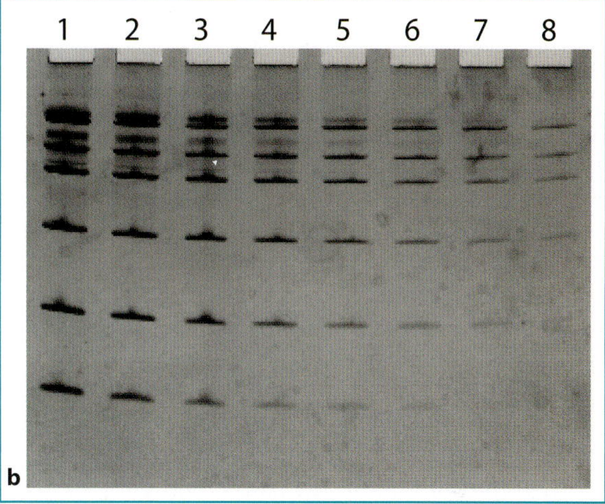

Low DNA mass ladder (Invitrogen, catalogue number. 10068-013)

	Concentration per 4 µl							
Fragment size (bp)	1	2	3	4	5	6	7	8
2000	100 ng	50	25	12.5	6.25	3.12	1.56	0.78
1200	60	30	15	7.5	3.75	1.87	0.93	0.46
800	40	20	10	5	2.5	1.25	0.62	0.31
400	20	10	5	2.5	1.25	0.62	0.31	0.15
200	10	5	2.5	1.25	0.62	0.31	0.15	0.07
100	5	2.5	1.25	0.62	0.31	0.15	0.07	0.04

Fig. 8.5 a, b (**a**) Example of a 2% agarose gel in 0.5× TAE. Running conditions: 90 V for 45 min. Sample: 8 dilutions (1:2) of marker in water; loading: 4 µl sample + 1 µl loading buffer. See table below for fragment size and concentrations. (**b**) Example of a 7% acrylamide gel in 1×TBE. Running conditions: 200 V for 40 min. Sample: 8 dilutions (1:2) of marker in water; loading: 4 µl sample+ 1 µl loading buffer. See table below for fragment size and concentrations. The DNA bands in lanes 5–8 are better visualised and are of higher definition than bands in the agarose gel, particularly for bands of smaller size. The separation of bands of similar length is also improved

viate the need for expensive visualisation equipment. Drawbacks include the use of toxic compounds (silver nitrate and acrylamide) and the longer time required to set up and run the gel compared to the agarose system. Please see the following pages for formulas and protocols.

8.9.2 Preparation of the SDS-Polyacrylamide Gel for Electrophoresis

Note: The recipes for the required solutions are given in Section 8.9.3.1.

Gel Rack Assembly (Fig. 8.6 b)

- Clean all of the components (gel glasses, spacer, gel comb, etc) with alcohol.
- Assemble the gel rack following the manufacturer's instructions.

Gel Preparation

Note: Volumes are for a vertical gel (dimensions 10 × 7 cm).

1. Mix the following reagents in a chemical hood:
   - *5 ml 7% Acrylamide solution (solution A)*
   - *20 μl APS (ammonium persulphate) 10% (solution B)*
   - *10 μl TEMED (N,N,N',N'-tetramethylethylenediamine; commercially available, ready to use. Sigma 7024 or Biorad 161-0800 or -0801)*
2. Load the solution in the assembled gel rack.
3. Place the gel comb in the gel and allow the gel to polymerise at room temperature for 20 min.

Sample Loading and Electrophoresis

1. Put the gel rack in the gel chamber. Fill with 1 × TBE buffer 1× (solution C, concentrate).
2. Mix the appropriate sample volume (usually 7 μl) with 3-4 μl of gel loading buffer.
3. Remove the gel comb.
4. Wash the wells with a syringe filled with TBE buffer.
5. Load the samples and the appropriate marker in the gel.
6. Close the gel chamber.
7. Turn on the power supplier and set the electrophoresis conditions (for a 7% acrylamide gel, usually 40 min; 200V, 400 mA).
8. Connect the gel chamber to the power supply and start the electrophoresis run.

Gel Staining

Note: The use of a shaking platform is advisable.

1. At the end of the run, turn off the power supplier and disassemble the gel chamber.
2. Open the gel rack and place the gel in a tank set on a shaking platform.
3. Cover the gel completely with fixative solution and shake for at least 5 min.
4. Recover the fixative solution and cover the gel completely with ddH_2O. Shake for 2 min.
5. Discard the water and cover the gel completely with silver nitrate solution. Shake for 5 min.
6. Recover the silver nitrate solution and cover the gel completely with ddH_2O. Shake for 2 min.
7. Discard the water and cover the gel completely with developing solution (solution F). Shake until bands become visible.
8. Discard the developing solution. Cover the gel completely with acetic acid solution (solution G).
9. Shake for at least 5 min.
10. Recover solution G and submerge the gel in ddH_2O. Shake for 2 min.
11. Wrap the gel in plastic wrap to make it easier to handle.

8.9.2.1 Solutions for SDS-polyacrylamide gel (SDS-PAGE) (Fig 8.6a)

Solution A: Acrylamide 7%

(per 500 ml solution)

- *50 ml 10× TBE*
- *87.5 ml Acrylamide/bis acrylamide 29:1 40%*
- *50 ml Glycerol 100%*

1. Add acrylamide 40%.
2. Add 10× TBE.
3. Add glycerol 100%.
4. Add 313 ml ddH_2O.
5. Store at room temperature (stable for maximum 1 year).

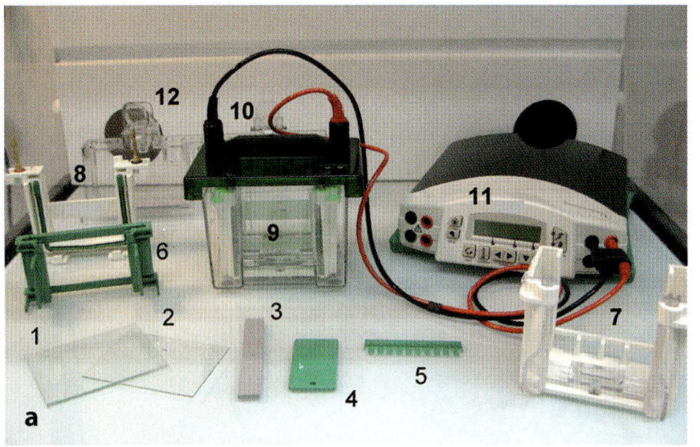

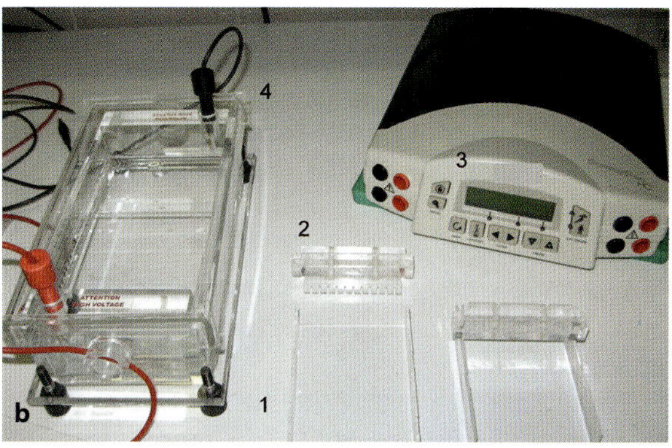

Fig. 8.6 a, b (**a**) *1* Spacer plates; *2* short plates; *3* casting stand gasket; *4* gel releaser; *5* comb; *6* casting frame; *7*, *8* clamping frame; *9* buffer tank; *10* lid; *11* power supply; *12* casting stand. (**b**) *1* UV-transparent tray; *2* comb; *3* power supply; *4* electrophoresis cell

Solution B: Ammonium Persulfate (APS) 10%

(per 10 ml solution)

- *1 g Ammonium persulphate*

1. Add ammonium persulphate to a 15-ml vial.
2. Dissolve ammonium persulfate in 10 ml ddH_2O.
3. Store at room temperature (stable for maximum 1 year).

Solution C: 10 × TBE

(per 1 l)

- *108 g Trizma base*
- *55 g Boric acid*
- *40 ml EDTA 0.5M, pH 8*

1. Weight Trizma base and boric acid.
2. Dissolve the salts with EDTA in ddH_2O and mix at room temperature,
3. Add ddH_2O to a final volume of 1 l.
4. Autoclave at 121°C for 20 min.
5. Store at +4°C (stable for maximum 1 year).

Solution D: Fixative

(per 1 l)

- *100 ml Methanol or ethanol absolute*
- *15 ml Nitric acid 70%*

1. Add 885 ml ddH_2O.
2. Add nitric acid 70%.
3. Add ethanol absolute or methanol.
4. Store at room temperature (stable for maximum 6 months).

Solution E: Silver Nitrate 0.4%

(per 500 ml)

- *2 g Silver nitrate*

1. Add silver nitrate.
2. Dissolve silver nitrate in 500 ml ddH_2O.
3. Filter the solution.
4. Store at room temperature.

Solution F: Developing Solution

(per 1 l)

- *30 g Sodium carbonate*
- *700 µl Formaldehyde 36%*

1. Dissolve sodium carbonate in 1 l ddH_2O.
2. Store at room temperature (stable for maximum 6 months).
3. At first use, add 700 µl formaldehyde 36%.

Solution G: Acetic Acid 5%

(per 1 l)

- *50 ml Glacial acetic acid, pure (100%)*

1. Add 950 ml ddH_2O to a bottle.
2. Add pure glacial acetic acid.
3. Store at room temperature.

8.9.3 Preparation of the Agarose Gel

Note: The recipes for the required solutions are given in Section 8.9.3.1.

Gel Preparation

1. Weight the correct amount of agarose powder (molecular biology grade) to obtain the expected concentration (%). The volume of the gel depends on the dimension of the tank and is calculated as follows:
   Gel tank width × Gel tank length × Gel thickness (0.5 cm)

Example 1:
   10 × 7.5 × 0.5 cm = 37.5 ml 1 × TAE = 0.375 g agarose for a 1% gel or 0.75 g agarose for a 2% gel.

Example 2:
   8.5 × 7 × 0.5 cm = 30 ml 1 × TAE = 0.3 g agarose for a 1% gel or 0.6 g agarose for a 2% gel.

2. Mix the agarose powder in the correct volume of TAE buffer (solution H).
3. Dissolve the agarose mixture in the microwave.
4. Water-cool the solution until it can be easily handled.
5. To stain gel by including ethidium bromide (EtBr): Add 1.8 µl of a 10 mg EtBr/ml (for a 30-ml gel) and mix. To stain gel by final immersion in EtBr, omit above step and see below.
6. Place the cooled solution in the gel tank prepared with the comb. Avoid the formation of air bubbles.
7. Allow the gel to solidify at room temperature.

Sample Loading and Electrophoresis

1. Extract the comb from the solidified gel and place the gel (with the tank) in an electrophoresis chamber.
2. Mix the loading sample (5–10 µl) with 3–4 µl of gel loading buffer (solution I).
3. Load the samples and the appropriate markers (solution J) in the gel.
4. Close the gel chamber.
5. Turn on the power supply and set up the electrophoresis conditions: 90–100 V, 30–40 min for fragments between 100 and 1000 bp in a 1% agarose gel.
6. At the end of electrophoresis:
   - If the gel was stained by the inclusion of EtBr, place the gel on a UV transilluminator.
   - Gel staining by final immersion in EtBr: immerse the gel in a tank with 0.5–1 µg EtBr/ml. Shake for 20 min. Recover the EtBr solution (can be re-used for up to 20 stainings) and submerge the gel in ddH_2O. Shake for 20 min. Place the gel with the tank on a UV transilluminator.

8.9.3.1 Solutions for Agarose Gel Electrophoresis

Solution A: 10 × TAE

(per 1 l)

- *48.4 g Trizma base*
- *11.4 ml Glacial acetic acid*
- *20 ml EDTA 0.5 M, pH 8*

1. Weigh Trizma base.
2. Add glacial acetic acid.
3. At room-temperature, add EDTA solution.
4. Add ddH_2O to a final volume of 1 l.
5. Autoclave at 121°C for 20 min.
6. Store at +4°C.

Solution B: Gel Loading Buffer 10 ×

(per 10 ml)

- *0.2 g Ficoll*
- *1 ml TBE 10×*
- *6 ml Glycerol 100%*
- *1 ml Bromophenol blue and xylene cyanol (0.5%)*

1. Weigh Ficoll in a 15-ml vial.
2. Add 1 ml TBE 10×.
3. Add 2 ml ddH_2O.
4. Dissolve the solution in a thermal bath at 37°C.
5. Add glycerol.
6. Add bromophenol blue and xylene cyanol solution.
7. Store at +4°C.

Solution C: Markers

(for 800 µl)

- *100 µl Markers (Roche)*
- *300 µl Gel loading buffer 10×*
- *400 µl TBE 1×*

1. Add markers to a 1.5-ml vial.
2. Add gel loading buffer.
3. Add TBE 1×-
4. Store at +4°C (stable for maximum 1 year).

8.10 Detection of Type A Influenza Virus by End-Point RT-PCR

8.10.1 Protocol 1

This protocol was developed at the Istituto Zooprofilattico Sperimentale delle Venezie (IZSVe) by adapting the primer set used in the rRT-PCR protocol developed by Spackman et al. (2002) to an end point, one-step RT-PCR. Laboratory observations indicated that this protocol is less sensitive than the original rRT-PCR, as expected. However, it can be easily and fruitfully used in laboratories that are not equipped with real-time platforms. Due to the small size of the amplicon (99 bp), the RT-PCR results should be visualised on silver-stained SDS-PAGE gels or on a 2–3% agarose gel. The Applied Biosystems GeneAmp Gold RNA PCR Core Kit (product number 4308207) has been used.

Primers

Forward M+25: AGA TGA GTC TTC TAA CCG AGG TCG
Reverse M-124: TGC AAA AAC ATC TTC AAG TCT CTG

Procedure

Reagent (concentrated stock solution)	Final concentration	1× Reaction (μl)
RNase-free water	/	3.2
PCR buffer (5×)	1×	5
$MgCl_2$ (25 mM)	2.5 mM	2.5
dNTPs mix (10 mM)	1 mM	2.5
DTT (100 mM)	10 mM	2.5
Primer M25F (5 pmol/μl)	0.3 μM	1.5
Primer M124R (5 pmol/μl)	0.3 μM	1.5
RNase inhibitor (20 U/μl)	10 U	0.5
Reverse Transcriptase 50 U/μl	15 U	0.3
Ampli Taq GOLD (5 U/μl)	2.5 U	0.5
Total volume Vortex the mix for a few seconds. Aliquot 20 μl in 0.2-ml PCR tubes.		20
Add RNA		5
Final reaction volume		25

Cycling Conditions

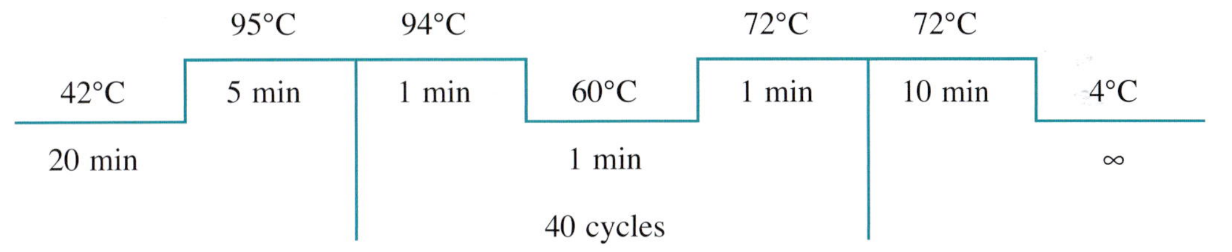

The amplicon is detected on a silver-stained 7% acrylamide gel or on a 2–3% agarose gel. The size of the expected amplified fragment is 99 bp.

8.10.2. Protocol 2

This protocol is a modification of the method developed by Fouchier et al. (2000) for the detection of type A influenza viruses in samples of human and animal origin, including birds. According to previous field investigation on swabs of avian origin, the relative sensitivity was 95.6% (CI_{95}=93.1–98.0) and the relative specificity 96.3% (CI_{95} = 94.4–98.1) compared to virus isolation (Cattoli et al. 2004). In some cases, unspecific bands are visualised on the gel (Martì et al. 2006). Careful examination of the gel and the use of proper controls and size-markers are therefore extremely important.The Applied Biosystems GeneAmp Gold RNA PCR Core Kit (product number 4308207) has been used.

Primers

Forward M52 C: 5'-CTT CTA ACC GAG GTC GAA ACG-3'
Reverse M253 R: 5'-AGG GCA TTT TGG ACA AAG/T CGT CTA-3'

Procedure

Reagent (concentrated stock solution)	Final concentration	1× Reaction (µl)
RNase-free water	/	4.7
PCR buffer (5×)	1×	5
$MgCl_2$ (25 mM)	2.5 mM	2.5
dNTPs mix (10 mM)	1 mM	2.5
DTT (100 mM)	10 mM	2.5
Primer M52 C (10 pmol/µl)	0.3 µM	0.75
Primer M253 R (10 pmol/µl)	0.3 µM	0.75
Rnase inhibitor (20 U/µl)	10 U	0.5
Reverse transcriptase (50 U/µl)	15 U	0.3
Ampli Taq GOLD (5 U/µl)	2.5 U	0.5
Total volume Vortex the mix for a few seconds. Aliquot 20 µl in 0.2-ml PCR tubes.		20
RNA		5
Final reaction volume		25

Cycling Conditions

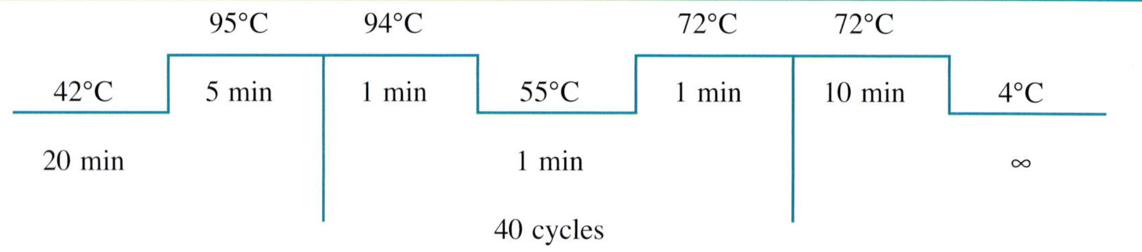

The amplicon is detected on a 2% agarose gel or 7% silver stained SDS-acrylamide gel. The size of the expected fragment is 244 bp.

8.11 Protocol for the Subtype-Specific Detection of H5 Avian Influenza Virus Using End-Point RT-PCR

8.11.1 Introduction

Two sensitive H5 PCR protocols have emerged from the AVIFLU European project (Slomka et al. 2007). In these protocols, amplicons are detected conventionally by agarose gel electrophoresis with EtBr staining or by SDS-PAGE with silver staining. Since both of the resulting amplicons span the HA cleavage site, sequencing can provide pathotype information, i.e. LPAI or HPAI. The two methods, which use different primer pairs, are known by their acronyms: H5KHA PCR and J3/B2a PCR. They are applicable to the detection of current Eurasian H5 isolates at clinical levels, including HPAI H5N1 isolates that have been circulating in poultry in the Far East since 2003 and the most recent H5N1 isolates circulating in Europe, the Middle East and Africa. In addition, other LPAI H5 strains isolated from European waterfowl within the past decade can be detected by either H5 PCR approach.

H5KHA PCR is highly sensitive, although in some cases possible specificity problems have been encountered. These include false-positives with non-H5 AI specimens and/or multiple bands similar in size to the predicted amplicon.

J3/B2a PCR is be less sensitive than the H5KHA PCR but can still successfully detect H5 amplicons from clinical specimens; however, difficulties in sequencing and pathotyping have been noted. Specificity appears to be better than that obtained with H5KHA PCR conditions.

Therefore, it may be appropriate to use both approaches for initial detection in a primary outbreak.

8.11.2 Protocol 1

Primers

H5-kha-1: CCT CCA GA**R** TAT GC**M** TA**Y** AAA ATT GTC
H5-kha-3: TAC CAA CCG TCT ACC AT**K** CC**Y** TG
Note the inclusion of degenerate nucleotides, indicated above in bold.

Procedure

Use the Qiagen OneStep RT-PCR kit (catalogue # 210212) to prepare 50-μl PCR volumes.

Reagent	Final concentration	Volume required for one reaction (μl)	Total (μl)
RNase-free water	/	28.8	
PCR buffer (5×, from Qiagen OneStep RT-PCR kit)	1×	10	
dNTPs mix (10mM each, from Qiagen kit)	0.4 mM each	2	
Primer H5-kha-1 (50 pmol/μl, 50 μM)	1 μM	1	
Primer H5-kha-3 (50 pmol/μl, 50 μM)	1 μM	1	
RNase inhibitor (40 U/μl, Promega)	8 U	0.2	
OneStep RT-PCR enzyme mix (Qiagen kit)		2	
Volume minus target		45	
Volume of extracted RNA		5	
Final reaction volume		50	

Cycling Conditions

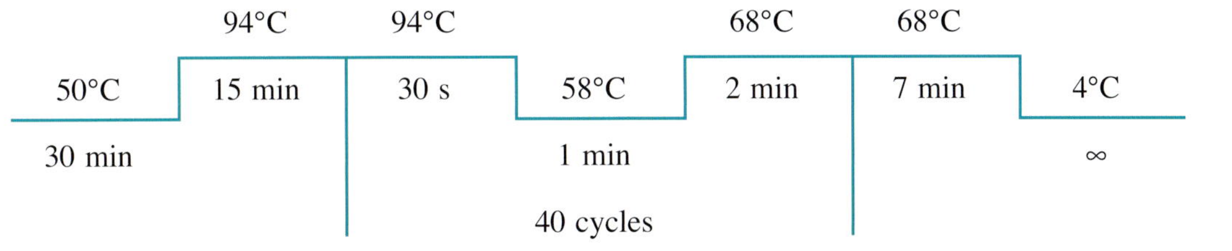

The amplicon is detected on a 2% agarose gel or 7% silver stained SDS-acrylamide gel. The size of the expected fragment is 300-320 bp.

8.11.3 Protocol 2

Primers

J3: GAT AAA TTC TAG CAT GCC ATT CC
B2a: TTT TGT CAA TGA TTG AGT TGA CCT TAT TGG

Procedure

Use the Qiagen OneStep RT-PCR kit (catalogue # 210212) to prepare 50-μl PCR volumes.

Reagent	Final concentration	Volume required for one reaction (μl)	Total (μl)
RNase-free water	/	28.8	
PCR buffer (5×, from Qiagen OneStep RT-PCR kit)	1×	10	
dNTPs mix (10 mM each, from Qiagen kit)	0.4 mM each	2	
Primer J3 (50 pmol/μl, 50 μM)	1 μM	1	
Primer B2a (50 pmol/μl, 50 μM)	1 μM	1	
RNase inhibitor 40 U/μl, Promega)	8 U	0.2	
OneStep RT-PCR enzyme mix (Qiagen kit)		2	
Volume minus target		45	
Volume of extracted RNA		5	
Final reaction volume		50	

Cycling Conditions

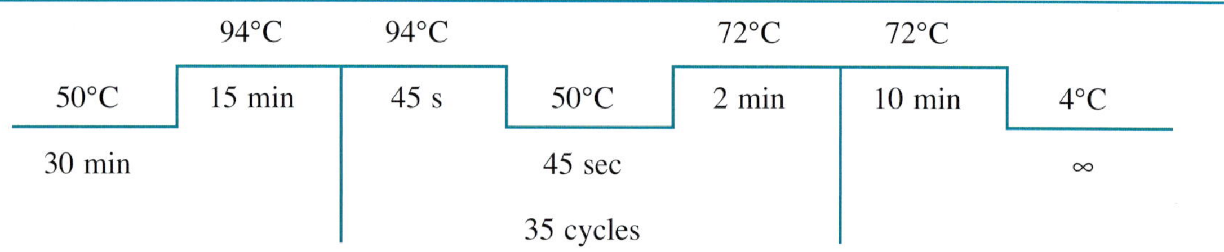

The amplicon is detected on a 2% agarose gel or 7% silver stained SDS-acrylamide gel. The size of the expected fragment is 300-320 bp

8.12 Protocol for the Subtype-Specific Detection of H7 Avian Influenza Virus Using End-Point RT-PCR

This two-step RT-PCR protocol has yielded good results in different tests conducted during the AVIFLU European project (Slomka et al. 2007). In this protocol, the amplicon is detected conventionally, either by agarose gel electrophoresis with EtBr staining or SDS-PAGE with silver staining. Since the amplicon spans the HA cleavage site, sequencing can provide pathotype information, i.e. LPAI or HPAI.

This two-step RT-PCR generates an amplicon of 200–220 bp.

8.12.1 Preparation of cDNA

Primers

GK 7.3 5'-ATG TCC GAG ATA TGT TAA GCA-3'
GK 7.4 5'-TTT GTA ATC TGC AGC AGT TC-3'

Procedure

Reagent	Final concentration	1 × Reaction (µl)
RNA/		10
Primer GK 7.3 (50 µM)	2.5 µM	1
Heat to 95°C for 2 min and then place the mixture immediately on ice.		
Add the following reagents:		
RNase-free water	/	3
M-MLV RT buffer (5×)	1×	4
dNTPs mix (10 mM)	0.5 mM each	1
RNase inhibitor (40 U/µl)	20 U	0.5
MMLV-RT (200 U/µl)	100 U	0.5
Final reaction volume		20

Cycling Conditions

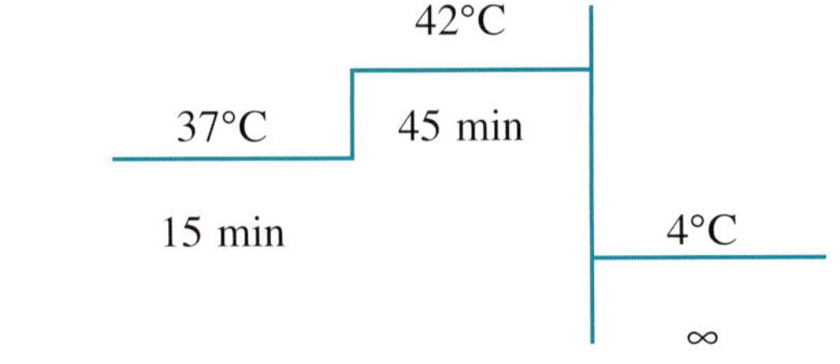

8.12.2 cDNA PCR Amplification

This protocol has been evaluated using the reagents contained in the AB Gene kit (catalogue #AB-0575/DC/LD/A).

Procedure

Reagent	Final concentration	1 × Reaction (μl)
RNase-free water	/	18
Reddy Mix PCR master mix with dNTPs (2×)	1×	25
Primer GK 7.3 (50 μM)	1 μM	1
Primer GK 7.4 (50 μM)	1 μM	1
Total volume		45
CDNA		5
Final reaction volume		50

Cycling Conditions

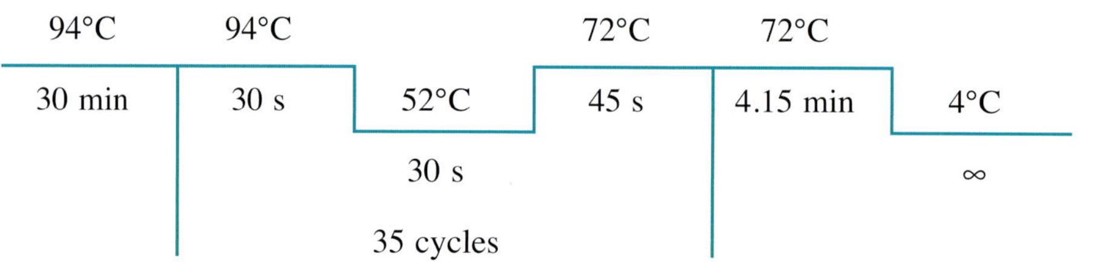

The amplicon is detected on a 2% agarose gel or 7% silver stained SDS-acrylamide gel. The size of the expected fragment is 200-220 bp

8.13 Detection of Type A Influenza Virus by Qualitative Real-Time PCR (M Gene)

8.13.1 Detection of Type A RNA by Real Time RT-PCR

The protocol uses the probe-primer set previously developed by Spackman et al. (2002). The basic procedure has been evaluated by other authors (Spackman et al. 2002; Slomka et al. 2007; Van Borm et al. 2007). In the one-step rRT-PCR protocol described below the QuantiTect Multiplex RT-PCR kit (Qiagen, product number 204643) has been used.

Primers

Forward M+25: AGA TGA GTC TTC TAA CCG AGG TCG
Reverse M-124: TGC AAA AAC ATC TTC AAG TCT CTG

Probe

FAM M+64 : FAM-5'-TCA GGC CCC CTC AAA GCC GA-3'–TAMRA

Procedure

Reagent	Final concentration	1 × Reaction (µl)
Probe FAM M+64 (1 µM)	100 nM	2.5
QuantiTect Multiplex RT-PCR master mix (2×)	1×	12.5
Primer M+25F (5 µM)	300 nM	1.5
Primer M-124R (5 µM)	300 nM	1.5
QuantiTect Multiplex RT mix	/	0.2
RNase-free water	/	1.8
Total volume Vortex the mix for a few seconds. Aliquot 20 µl per tube.		20
RNA		5
Final reaction volume		25

Cycling Conditions

This protocol was evaluated on AB7300 (Applied Biosystems) and on Rotorgene 6000 (Corbett) real-time platforms. Laboratories using different instrumentation platforms should first critically and carefully examine these cycling conditions, as they may *not* perform optimally on different instruments.

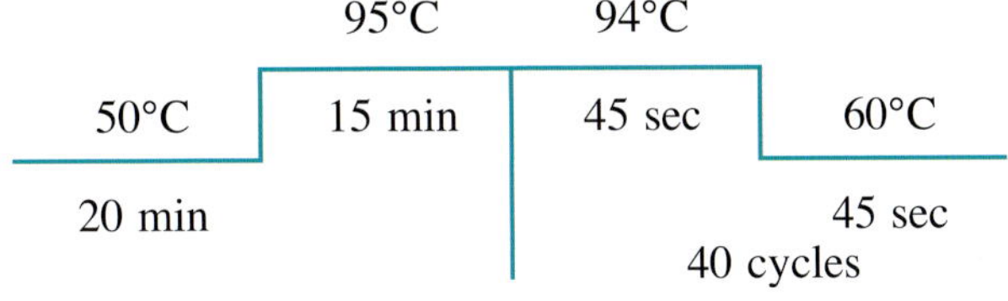

8.14 Detection of Influenza Virus of H5 Haemagglutinin Subtype by Qualitative One-Step Real Time RT-PCR

This protocol was developed and validated at the OIE/FAO Reference Laboratory in Italy (IZSVe, Legnaro–Padova). The procedure can be coupled with H7 and H9 detection within the same real-time PCR run, thus providing a fast and sensitive method for the detection of notifiable AI subtypes in poultry (Monne et al. 2008). The protocol has been validated on a wide variety of H5 isolates of the Eurasian lineage, including the recent H5 LPAI viruses circulating in poultry and wild birds as well as the H5N1 HPAI viruses circulating in Eastern and Central Asia, the Middle East, Europe and Africa.

The protocol was developed using the QuantiTect Multiplex RT-PCR kit (Qiagen product number 204643).

Degenerate nucleotides are indicated in bold.

Primers

Forward H5 F: TTA TTC AAC AGT GGC GAG
Reverse H5NE-R: CCA **K**AA AGA TAG ACC AGC

Probe

FAM H5: FAM-5'-CCC TAG CAC TGG CAA TCA TG-3'-TAMRA

Procedure

Reagent (concentrated stock solution)	Final concentration	1 × Reaction (µl)
Probe FAM H5 (1 µM)	150 nM	3.75
QuantiTect Multiplex RT-PCR master mix (2×)	1×	12.5
Primer H5F (5 µM)	300 nM	1.5
Primer H5NE-R (5 µM)	300 nM	1.5
QuantiTect Multiplex RT mix		0.2
RNase-free water	/	0.55
Total volume Vortex the mix for a few seconds. Aliquot 20 µl per tube.		20
RNA		5
Final reaction volume		25

Cycling Conditions

This protocol was evaluated on AB7300 (Applied Biosystems) and on Rotorgene 6000 (Corbett) real-time platforms. Laboratories using different instrumentation platforms should first critically and carefully examine these cycling conditions, as they may not perform optimally on different instruments.

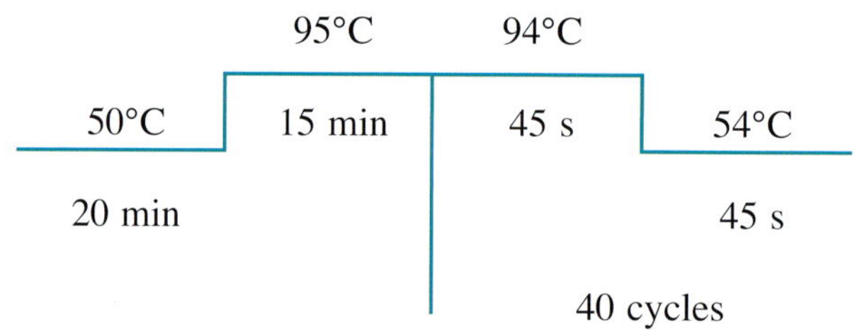

8.15 Alternative Protocol for the Detection of H5 Avian Influenza Virus by Real-Time RT-PCR

Primers and Probe

The primers and probe are a modification of those designed originally by Spackman et al. (2002). Modified sequences are as follows (Slomka et al. 2007b):
H5LH1: ACA TAT GAC TAC CCA CA**R** TAT TCA G
H5RH1: AGA CCA GCT A**Y**C ATG ATT GC
H5PRO: FAM-TC**W** ACA GTG GCG AGT TCC CTA GCA-TAMRA
Degenerate nucleotides are indicated in bold. The above primers and probe include modifications to detect current H5 (including HPAI viruses) Eurasian avian influenza viruses.

H5 Real-Time PCR Master Mix

This is based on the Methods section of Spackman et al. (2002). The protocol has been applied using the QuantiTect Multiplex RT-PCR kit (Qiagen product number 204643).

Procedure

Reagent	Final concentration	1 × Reaction (µl)
RNase-free water	/	4.95
QuantiTect Multiplex RT-PCR master mix (2×)	1 ×	12.5
Primer H5LH1 (5 µM)	400 nM	2
Primer H5RH1 (5 µM)	400 nM	2
Probe FAM PRO-H5 (6 µM)	300 nM	1.25
RNasin (40 U/µl)	4 U	0.1
QuantiTect Multiplex RT mix		0.2
Total volume Vortex the mix for a few seconds. Aliquot 23 µl per tube.		23
RNA		2
Final reaction volume		25

Cycling Conditions

This protocol was evaluated on AB7300 (Applied Biosystems) and on Rotorgene 6000 (Corbett) real-time platforms. Laboratories using different instrumentation platforms should first critically and carefully examine these cycling conditions, as they may *not* perform optimally on different instruments.

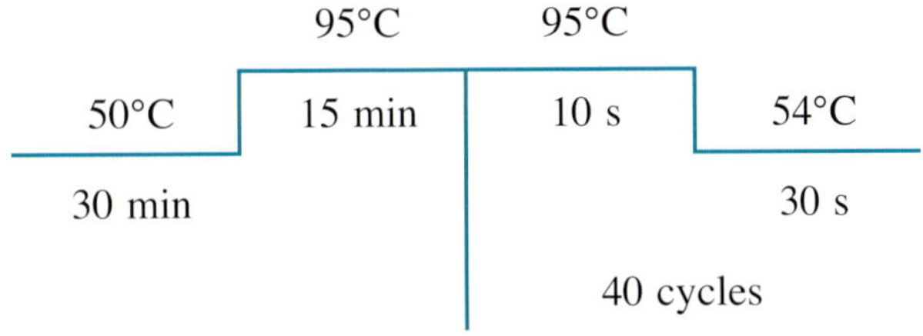

8.16 Detection of Influenza Virus of H7 Haemagglutinin Subtype by Qualitative Real Time PCR (IZSVe Protocol)

This protocol was developed and validated at the OIE/FAO Reference Laboratory in Italy (IZSVe, Legnaro–Padova). The procedure can be coupled with H5 and H9 detection within the same rPCR run, providing a fast and sensitive method for the detection of notifiable AI subtypes in poultry (Monne et al. 2008). The protocol has been validated on a wide variety of H7 isolates of the Eurasian lineage, including the recent H7 LPAI viruses circulating in poultry and wild birds in Europe and Africa as well as the H7 HPAI viruses that caused the recent outbreaks in Europe (e.g. Italy H7N1, The Netherlands H7N7).

The protocol was developed using the QuantiTect Multiplex RT-PCR kit (Qiagen Cod. 204643).

Primers

Forward **H7 F**: TTT GGT TTA GCT TCG GG
Reverse **H7-degR**: GAA GA**M** AAG GC**Y** CAT TG
Degenerate nucleotides are indicated in bold.

Probe

VIC **H7**: VIC-5'-CAT CAT GTT TCA TAC TTC TGG CCA T-3'-TAMRA

Procedure

Reagent	Final concentration	1 × Reaction (µl)
Probe VIC H7 (1 µM)	150 nM	3.75
QuantiTect Multiplex RT-PCR master mix (2×)	1×	12.5
Primer H7F (10 µM)	300 nM	0.75
Primer H7-degR (10 µM)	900 nM	2.25
QuantiTect Multiplex RT mix		0.2
RNase-free water	/	0.55
Total volume Vortex the mix for a few seconds Aliquote 20 µl per tube		20
RNA		5
Final reaction volume		25

Cycling Conditions

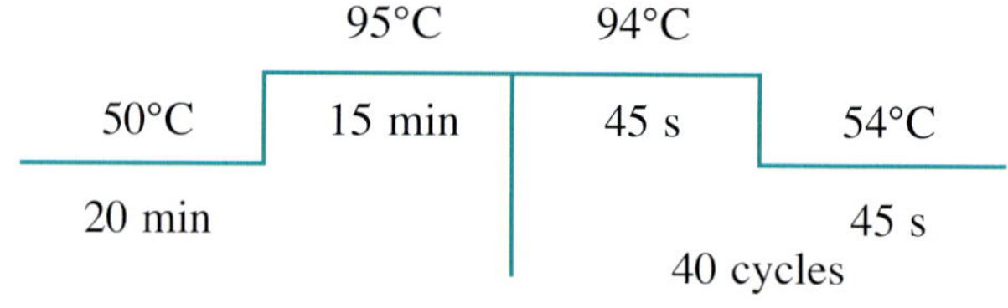

References

Cattoli G, Drago A, Maniero S et al (2004) Comparison of three rapid detection systems for type A influenza virus on tracheal swabs of experimentally and naturally infected birds. Avian Pathol 33(4):432-437

EC (European Commission) (2006) Commission Decision 2006/437/EC of 4 August 2006 approving a Diagnostic Manual for avian influenza as provided for in Council Directive 2005/94/EC (notified under document number C(2006) 3477) Available at: http://eur-lex.europa.eu/LexUriServ/site/en/oj/ 2006/l_237/ l_23720060831en00010027.pdf

Fouchier RA, Bestebroer TM, Herfst S et al (2000) Detection of influenza A viruses from different species by PCR amplification of conserved sequences in the matrix gene. J Clin Microbiol 38(11):4096-4101

Martì NB, Del Pozo ES, Casals AA et al (2006) False-positive results obtained by following a commonly used reverse transcription-PCR protocol for detection of Influenza A virus. J Clin Microbiol 44(10):3845

Monne I, Ormelli S, Salviato A et al (2008) Development and validation of a one-step real-time PCR assay for simultaneous detection of subtype H5, H7 and H9 avian influenza viruses. J Clin Microbiol 46(5):1769-1773

OIE (World Organization for Animal Health) (2004) Highly Pathogenic Avian Influenza. In: Manual of diagnostic tests and vaccines for terrestrial animals, 5th Edition, Office International des Epizooties, Paris, France, 258

Slomka MJ, Coward VJ, Banks J et al (2007) Identification of sensitive and specific avian influenza polymerase chain reaction methods through blind ring trials organized in the European Union. Avian Dis 51(1 Suppl):227-234

Slomka MJ, Pavlidis T, Banks J et al (2007b) Validated H5 Eurasian real-time reverse transcriptase-polymerase chain reaction and its application in H5N1 outbreaks in 2005-2006. Avian Dis 51(1 Suppl):373-377

Spackman E, Senne DA, Myers TJ et al (2002) Development of a real-time reverse transcriptase PCR assay for type A influenza virus and the avian H5 and H7 hemagglutinin subtypes. J Clin Microbiol 40(9):3256-3260

Van Borm S, Steensels M, Ferreira HL et al (2007) A universal avian endogenous real-time reverse transcriptase-polymerase chain reaction control and its application to avian influenza diagnosis and quantification. Avian Dis 51(1 Suppl):213-220

9 Clinical Traits and Pathology of Newcastle Disease Infection and Guidelines for Farm Visit and Differential Diagnosis

Calogero Terregino and Ilaria Capua

9.1 Introduction

The main body of evidence regarding the clinical aspects of avian paramyxovirus type 1 (APMV-1) infection has been collected from poultry, mainly chickens. Based on the occurrence and severity of clinical manifestations, Beard and Hanson (1984) identified five viral pathotypes (Table 9.1). The clinical manifestations of this disease are highly variable and there are no lesions or signs that can be considered pathognomonic (McFerran and McCracken 1988).

Clinical signs produced by the same virus are influenced by numerous factors, including the species infected, the age and the production status or health of the host, especially in the presence of co-infections with other viruses, bacteria or parasites. In addition, vaccination against Newcastle disease (ND) virus infection is carried out globally. Consequently, clinical signs may also vary depending on the level of immunity to the virus, which may be passively derived from maternal antibodies or actively induced by vaccination. Several factors influence the immune status to ND following vaccination, including underlying immunosuppressive diseases, quality and type of vaccine, and number of administrations. The clinical manifestations of infection, even with highly virulent viruses, may therefore not be as overt as described and illustrated in this chapter.

9.2 Chickens and Turkeys

Velogenic ND is an acute condition that affects birds of all ages and categories. In naïve birds, ND is often characterised by a sudden onset of clinical manifestations. Some birds die peracutely, prior to the onset of clinical signs, whereas others show more general signs of disease such as anorexia, ruffled feathers and dropped wings. In laying birds, the most pronounced sign is a marked drop in egg production or a complete cessation of egg laying. Eggs are often misshapen, with thin shells and watery albumen. Depending on the tropism of the strain involved, clinical manifestations may occur predominantly in the gastrointestinal tract (velogenic viscerotropic, VVND) leading to a severe enteritis mainly characterised by diarrhoea, which is often green in colour. In contrast, velogenic neurotropic forms are dominated by res-

Table 9.1 Clinical course of avian paramyxovirus type 1 (APMV-1) infection in chickens (*Gallus gallus* var. dom.) (Modified from Beard and Hanson. 1984)

	Velogenic		Mesogenic	Lentogenic	Asymptomatic
	Viscerotropic	Neurotropic			Enteric
Diarrhoea	+++	–	–	–	–
Respiratory distress	–	+++	++	(+)	–
Central nervous system signs	(++)	+++	(++)	–	–
Drop in egg production	+++	+++	++	(+)	–
Morbidity	+++	+++	++	(+)	–
Mortality	+++	++	+	(+)	–

Severity of signs observed: +++ severe, ++ intermediate, + mild, ( ) clinical signs only in compromised or young birds

I. Capua, D.J. Alexander (eds.) *Avian Influenza and Newcastle Disease,* piratory distress, which is followed by central nervous system disorders (Figs. 9.1–9.6). For both velogenic forms, flock mortality levels in fully susceptible birds may be as high as 90–100%. Some velogenic viruses cause a less severe disease in turkeys than in chickens.

In contrast, clinical manifestations of infection with mesogenic viruses are strongly dependent on the age of the infected animals. In young birds, morbidity within a flock can be as high as 100%, while in adult healthy chickens it ranges between 5% and, exceptionally, 50%. The main clinical signs of infection with a mesogenic virus are a drop in egg production, poor egg quality (shell-less or soft-shelled eggs, off-coloured eggs; (Figs. 9.7, 9.8) and decreased feed consumption. However, most of the so-called mesogenic viruses are not naturally occurring; rather, they are velogenic viruses that have been attenuated by a variety of methods in the laboratory. There is evidence that some of these viruses may revert to the virulent phenotype after passage in chickens.

Lentogenic pathotypes such as B1 and La Sota are usually apathogenic in adult birds and are used as live vaccines. The same viruses, if administered to 1- to 7-day-old chicks, can cause reduced food intake and respiratory distress, the latter characterised by sneezing and snicking (Alexander 2003).

In addition to the age of the infected animals, variability in the clinical course of ND may arise due to co-infections with other microorganisms, in-

Fig. 9.1 Layer hens, naturally infected with Newcastle disease (ND) virus, velogenic neurotropic pathotype, exhibiting nervous signs

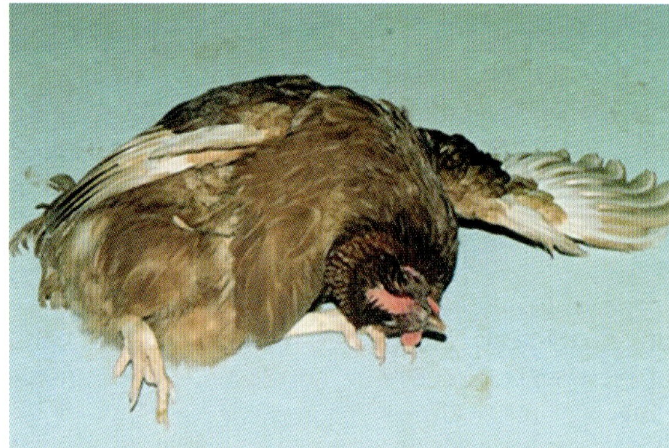

Fig. 9.2 Layer hen, naturally infected with ND virus, velogenic neurotropic pathotype, exhibiting serious nervous signs with torticollis and paresis

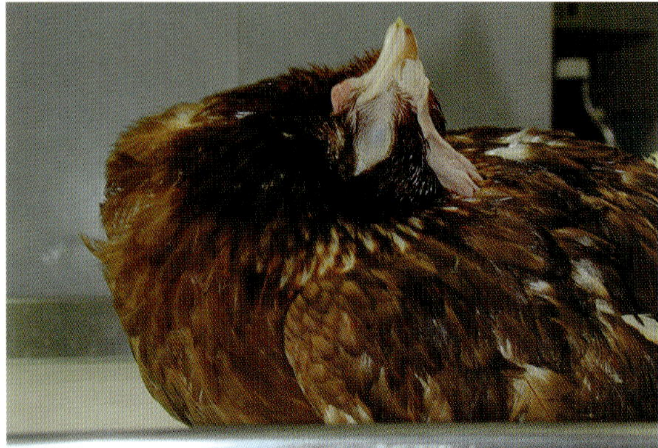

Fig. 9.3 Layer hen, naturally infected with ND virus, velogenic neurotropic pathotype, exhibiting torticollis

Fig. 9.4 Caged layer hens, naturally infected with ND virus, velogenic neurotropic pathotype, exhibiting nervous signs with paresis

Fig. 9.5 Layer hen, naturally infected with ND virus, velogenic neurotropic pathotype, exhibiting serious nervous signs with incoordination of muscular movement

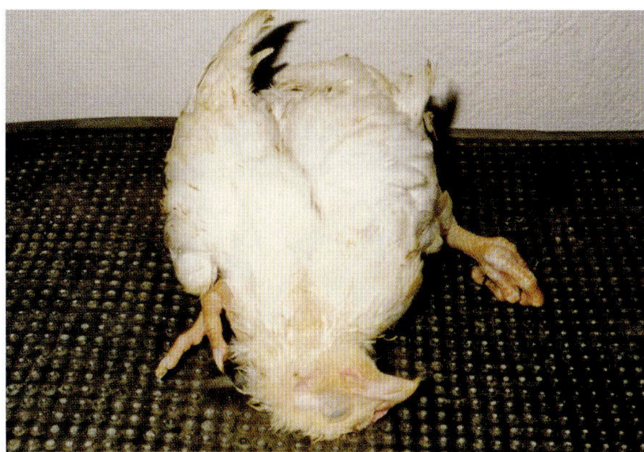

Fig. 9.6 Chicken experimentally infected with ND virus, velogenic neurotropic0 pathotype, exhibiting torticollis and toe paralysis. (Courtesy of Dr. Zenon Minta)

Fig. 9.7 Soft-shelled, irregular-shaped and off-coloured eggs produced by hens affected by ND

Fig. 9.8 Discoloured eggs produced by hens affected by ND (*right*)

cluding *Mycoplasma*, *Escherichia coli* or other APMV viral pathogens will lead to more pronounced clinical signs (Gross 1961; Kim et al. 1978, Nakamura et al. 1994). The animal's immune status also greatly influences the outcome of ND virus infection. In immunocompromised animals, for example, as a result of infection with viruses such as infectious bursal disease virus (which produces Gumboro disease), chicken anaemia virus, haemorrhagic enteritis virus or Marek's disease virus, infection with ND strains of low virulence can lead to overt clinical disease with subsequent economic losses in terms of reduced performance and mortality.

9.3 Ostriches

Ostriches (*Struthio camelus*) are considered to be moderately susceptible to ND and outbreaks have been reported in zoo and farmed ostriches (Alexander 2000). Clinical signs of disease include inappetence, apathy, ataxia and torticollis. The duration of ND in adult animals is estimated to be 3–16 days (Kauker and Siegert 1957; Verwoerd 1995). Clinical manifestations are predominant in younger birds between 5 and 9 months of age. Clinical signs include those involving the nervous system, such as atonic paralysis of the neck, torticollis, rhythmic twitches of the muscles of the back, oedema of the head and total paralysis, leading to death in approximately 30% of infected animals (Samberg et al. 1989). Different clinical manifestations, such as respiratory distress due to haemorrhagic tracheitis, have been reported in ostrich chicks reared indoors (Huchzermeyer 1996).

9.4 Game Birds

Reports on clinical manifestations following natural infection of game birds are few, although partridges and pheasants are highly susceptible to ND (Aldous and Alexander 2008). In pheasants, clinical signs reported in natural outbreaks are highly variable and resemble those seen in chickens. The disease can appear in an acute form, with sudden onset, nervous signs (incoordination, head shaking) and high mortality, or as a mild disease with respiratory distress, blindness and ataxia as the only detectable clinical signs. There is a subclinical (asymptomatic) form of ND as well as many intermediate forms. The clinical signs include drooping wings and depression, lack of appetite, respiratory distress with beak gaping, coughing, sneezing, gurgling and rattling and yellowish-green diarrhoea. In laying flocks, a sudden drop in egg production and a high proportion of eggs laid with abnormal (soft) shells is often an early sign of disease. Young birds are particularly susceptible and mortality can be extremely high, with survivors often exhibiting permanent nervous signs.

9.5 Ducks, Geese and Swans

Some waterfowl are known to be highly resistant to the clinical manifestations of ND, although they are susceptible to infection. On rare occasions, infection has been associated with a mild to severe clinical condition. Outbreaks in geese flocks have been characterised by signs ranging from ruffled feathers or mild depression to severe systemic infection with anorexia, white diarrhoea, ocular and nasal discharges and in some birds red and oedematous eyelids. The disease spreads very rapidly, with high fatality. Some birds die overnight, others shortly after the appearance of signs and others after a relatively prolonged course (between 3 and 12 days post-infection).

Natural infections of domestic ducks resulting in clinical manifestations are exceptional findings that have been associated with mortality and acute nervous signs (Kingston et al. 1978). Natural infections leading to clinical disease in wild Anatidae have been documented only rarely. Bozorgmehri-Fard and Keyvanfar (1979) reported rapid deaths in captured teals (*Anas crecca*) from which APMV-1 was isolated. Estudillo (1972) described respiratory, enteric and central nervous system signs in a mute swan (*Cygnus olor*), central nervous signs in a trumpeter swan (*Cygnus buccinator*) and respiratory and central nervous signs in a snow goose (*Chen caerulescens*) and a Canada goose (*Branta canadensis*), after natural infection with APMV-1.

Experimental inoculations of adult wild mallard ducks with a highly virulent form of ND virus isolated from chickens resulted in onset of clinical signs 2 days after inoculation (Friend and Trainer 1972; Friend and Franson, 1999). Initially, the mallards lay on their sternum with their legs slightly extended to the side. As the disease progressed, they were unable to rise when approached, lying on their sides and then exhibiting a pedalling motion with both legs in vain attempts to escape. Breathing in these birds was rapid and deep. Other mallards were unable to hold their heads erect. By day 4, torticollis and wing droop began to appear, followed by paralysis of one or both legs. Muscular tremors also became increasingly noticeable at this time.

9.6 Pigeons (*Columba livia*)

Clinical manifestations of ND virus infection in pigeons vary greatly, depending above all on the age and immune status of the bird and the pathogenicity of the viral strain causing infection. Pigeons and *Columbiformes* in general are mainly infected with the pigeon variant of APMV-1, known as pigeon paramyxovirus type 1 (PPMV-1). This virus is currently endemic in the pigeon population throughout the world. PPMV-1 infection should always be kept in mind when excess mortality or mild to severe illness affects pigeon flocks. Juvenile pigeons are very susceptible to infection; at times, the morbidity and mortality in young bird flocks may reach 100%, with nervous signs being dominant. In contrast, affected adults may completely recover after 10–14 day of illness. In adult birds morbidity is variable but often below 10%, with death occurring as a result of chronic disease and emaciation. In addition, subclinical infection seems to be common, contributing to the spread of the virus.

The incubation period is 7–14 days (Alexander et al. 1984), with virus being shed in the faeces as early as 2 day post-infection (Alexander and Parsons 1984) Infection may spread directly during the incubation period or indirectly through contaminated fomites (cages, transport vehicles, exhibitions).

Clinical signs in naïve birds resemble those of the neurotropic form of ND in chickens. Initially, affected birds exhibit general signs, such as poor body condition, reduced feed intake, increased drinking (polydipsia), increased excretion of urea (polyuria), lethargy, inability to fly and ruffled feathers. During the next few days, clinical signs of neuronal disorders appear, such as incoordination, abnormal gait, tremors, paresis of the legs and/or wings, head tilting, torticollis and greenish diarrhoea.

9.7 Pet Birds

Clinical manifestation of APMV-1 infection in parrots varies greatly, depending on the species and on the virus involved in the outbreak (Gerlach 1994; Kaleta and Baldhof 1988). The incubation period is generally 3–6 days, but may be as long as 14 days. Mortality can reach 100% in certain outbreaks, but may be as low as 22%, in others. Morbidity, mortality and clinical manifestations are highly variable between different species (Erickson et al. 1977b; Sallerman 1973). Clinical signs may be non-specific, such as depression, apathy and ruffled feathers, watery greenish diarrhoea, polyuria and, later, evidence of central nervous system involvement. Some pet birds may recover from infection and have been shown to shed virus for a prolonged period of time—for some psittacines, even more than a year (Erickson 1977a).

Canaries (*Serinus canarius*) were found to be resistant to clinical disease following experimental infection with high doses of a strain highly virulent for chickens (Terregino et al. 2004). Clinical signs, when present, may include severe depression (Fig. 9.9) and nervous signs (incoordination, torticollis, lying on back) prior to death (Fig. 9.10).

9.8 Gross Lesions

As with clinical signs, the gross lesions and the organs involved in birds infected with ND virus depend on the pathotype of the infecting virus, the host and all the other factors that determine the severity of the disease. No pathognomonic lesions are associated with any particular disease form. Gross lesions may also be absent.

Carcases of birds dying as a result of virulent ND virus usually have a fevered, dehydrated appearance. In acute forms of infection caused by these viruses the

Fig. 9.9 Canary (*Serinus canarius*) affected by virulent ND (VND) virus, during the acute phase of the disease, showing depression and ruffled feathers

Fig. 9.10 Canary (*Serinus canarius*) affected by VND virus, showing nervous signs

only clear lesions may be diffuse haemorrhages. Haemorrhagic lesions associated with virulent ND virus infection are often located in the intestine, most prominently in the mucosa of the proventriculus (Fig. 9.11), caeca and small intestine. Sometimes necrotic foci are observed in the pancreas (Figs. 9.12, 9.13). Petechial and small ecchymotic haemorrhages are often present on the mucosa of the proventriculus, near the base of the papillae, and concentrated around the posterior and anterior orifices.

Spleen, Peyer's patches, caecal tonsils and other focal aggregations (Fig. 9.14) of lymphoid tissue in the gut wall usually are markedly involved and are responsible for the term viscerotropic, applied to this form of ND (Figs. 9.15–9.17). These areas progressively become oedematous, haemorrhagic, necrotic

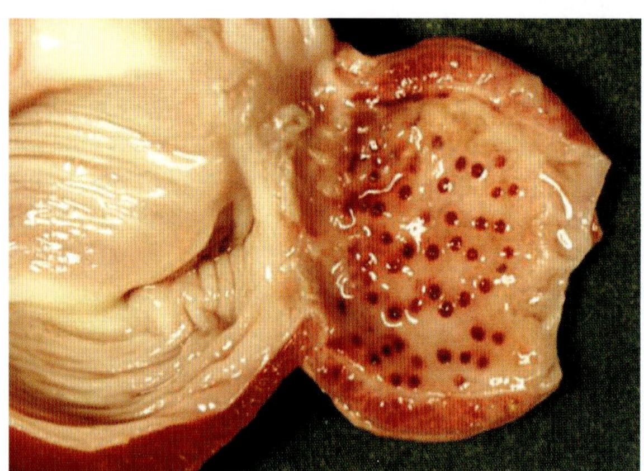

Fig. 9.11 Petechial hemorrhages around the ducts of the proventricular glandular region. (Courtesy of A. H. Zahdeh J.)

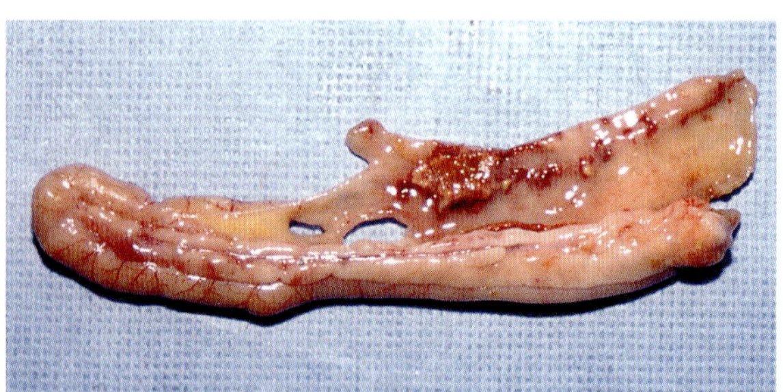

Fig. 9.12 Guinea fowl, naturally infected with VND virus, exhibiting pancreatitis and necrotic-haemorrhagic lesions in the intestine

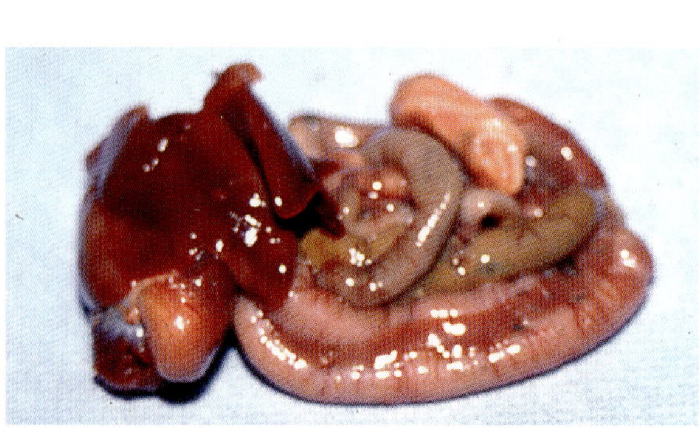

Fig. 9.13 Pheasant, naturally infected with VND virus, exhibiting pancreatitis and duodenitis

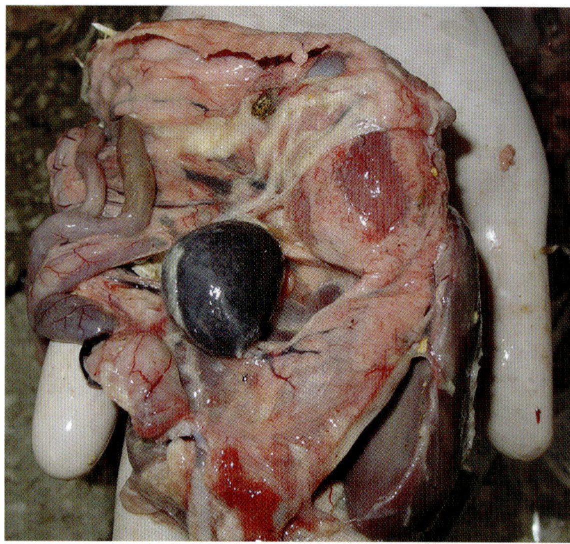

Fig. 9.14 Broiler, naturally infected with VND virus, exhibiting splenic enlargement and mottling due to necrosis. (Courtesy of Corrie Brown)

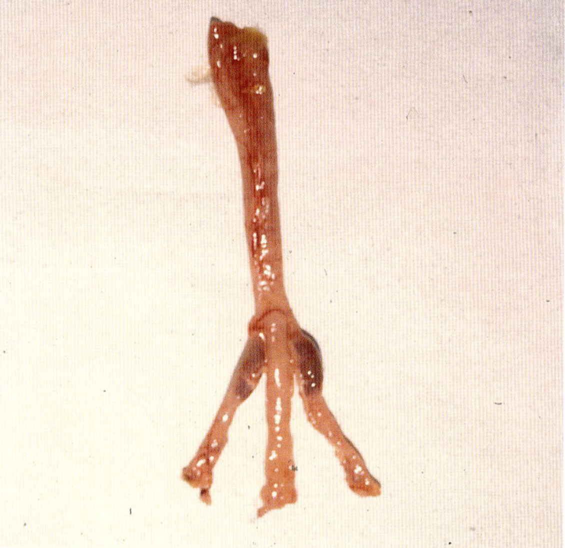

Fig. 9.15 Chicken, naturally infected with VND virus, exhibiting haemorrhagic lesions in caecal tonsils as seen through the serosal wall

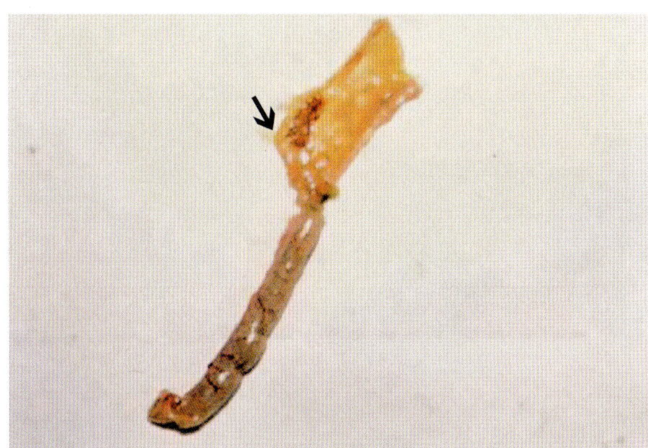

Fig. 9.16 Chicken, naturally infected with VND virus, exhibiting necrotic-haemorrhagic lesion in lymphatic intestinal tissue

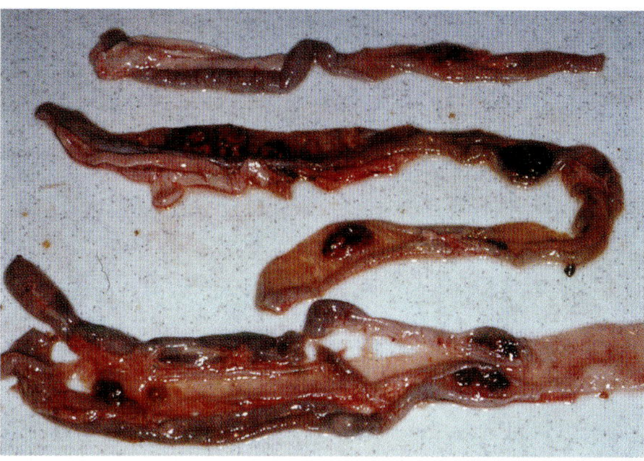

Fig. 9.17 Chicken, experimentally infected with VND virus, exhibiting necrotic-haemorrhagic lesion in lymphatic intestinal tissue. (Courtesy of Zenon Minta)

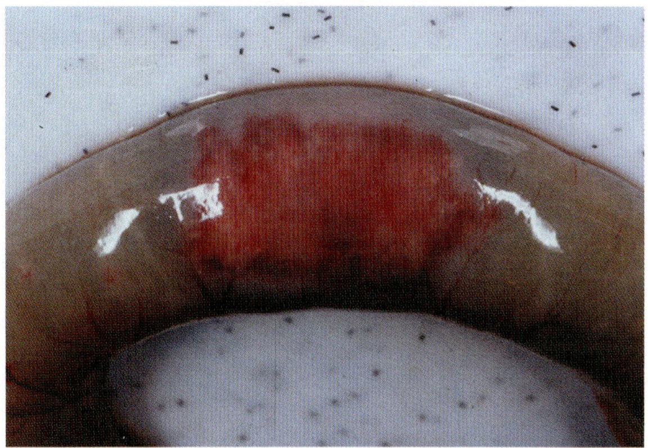

Fig. 9.18 Chicken, experimentally infected with VND virus, exhibiting a necrotic-haemorrhagic lesion in lymphatic intestinal tissue visible through the serosal wall. (Courtesy of Zenon Minta)

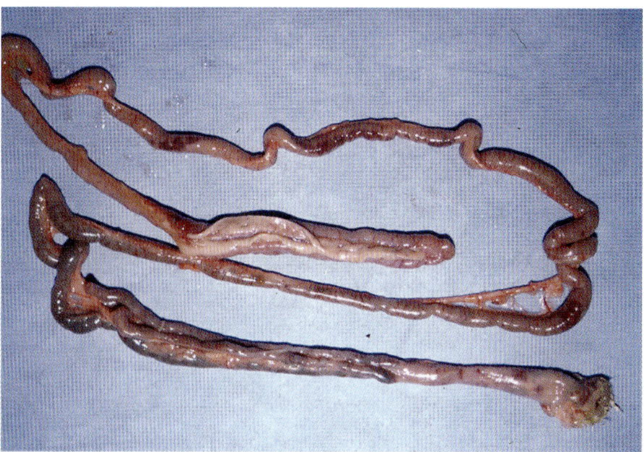

Fig. 9.19 Guinea fowl, naturally infected with VND virus, exhibiting a necrotic-haemorrhagic lesions of lymphatic intestinal tissue through the serosal wall

and ulcerative. In chickens that have died from VVND, lymphoid areas can often be observed without opening the gut (Figs. 9.18–9.20).

Ovaries may be oedematous, haemorrhagic or degenerated. Yolk peritonitis (Fig. 9.21) can frequently be observed in layers as a result of VVND; rough, misshapen eggs are typically laid by recovering hens.

Generally, gross lesions are not observed in the central nervous system of birds infected with ND virus, regardless of the pathotype and species. In case of disease manifestation in the respiratory tract, gross pathologic changes consist predominantly of mucosal haemorrhage and marked congestion of the trachea and lung (Figs. 9.22, 9.23). Air sacculitis may be present even after infection with low virulence strains, facilitating secondary bacterial infection with thickening of the air sacs and catarrhal or caseous exudates.

9.9 Guidelines for Farm Visits

It is important to obtain detailed clinical histories and field observations when suspected outbreaks of ND are investigated. The information should include the onset and type of clinical signs, mortality and morbidity, age and breed of the birds and the management procedures, including vaccination history (see also 6.4.2 and Annex 2).

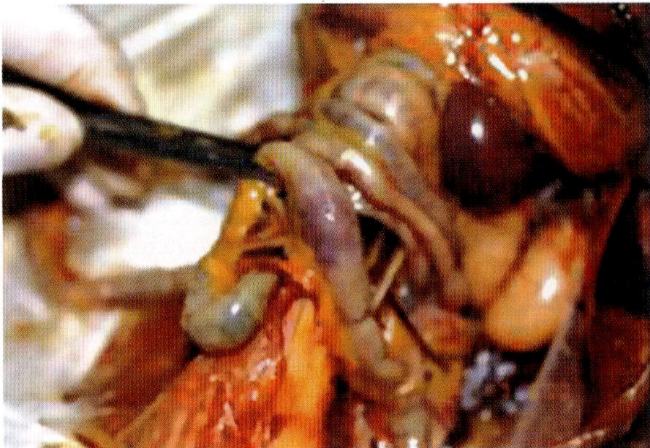

Fig. 9.20 Layer hen, naturally infected with VND virus, exhibiting necrotic-haemorrhagic lesions of lymphatic intestinal tissue visible through serosal wall

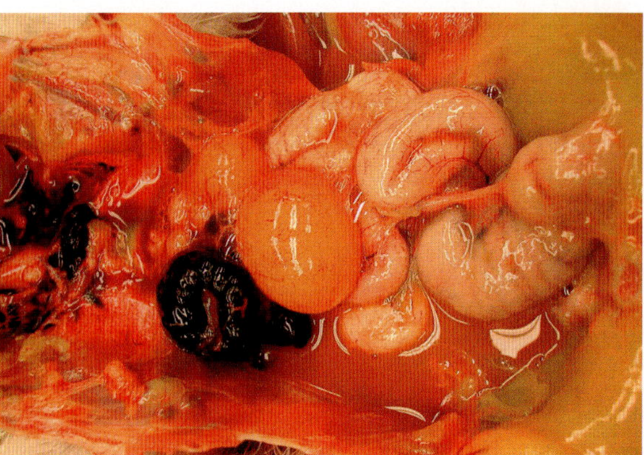

Fig. 9.21 Layer hen naturally infected with ND, exhibiting egg-yolk peritonitis. (Courtesy of Desiree Jansson)

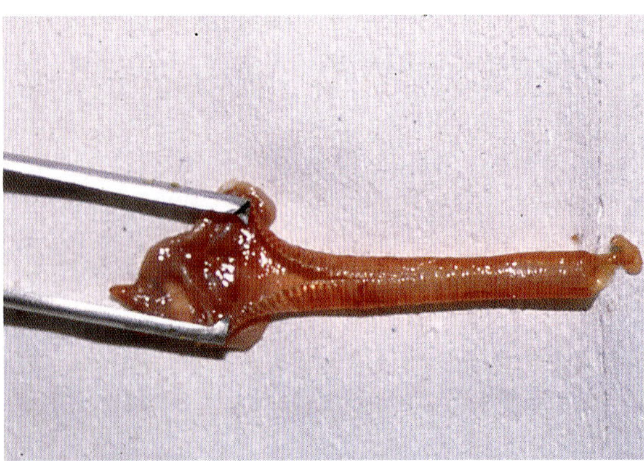

Fig. 9.22 Pheasant, naturally infected with VND virus, exhibiting haemorrhages in the larynx

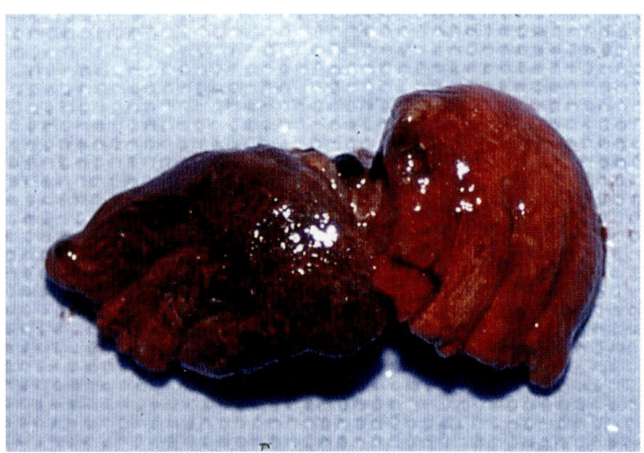

Fig. 9.23 Guinea fowl, naturally infected with VND virus, exhibiting bilateral pneumonia with haemorrhages

9.9.1 Summary of Main Clinical Signs Associated with ND Infection

The incubation period of ND after natural exposure has been reported to vary from 2 to 15 days (average 5–6). The speed with which signs appear, if at all, is variable, depend on the infecting virus, the host species and its age and immune status, infection with other organisms, environmental conditions, types of poultry farms, the route of exposure and the dose. Clinical signs in the field are variable and reflect the virulence and tropism of the infecting virus as well as the host species and its age and immune status (McFerran and McCracken 1988). In unvaccinated birds infected with extremely virulent viruses, ND is suspected in any flock in which sudden deaths or high mortality follow severe depression, inappetence, respiratory or enteric signs and a drastic decline in egg production. The disease may appear suddenly; mortality may be high even in the absence of other clinical signs. In outbreaks in chickens infected with the velogenic viscerotropic pathotype (VVND), clinical signs often begin with listlessness, increased respiration and weakness and end with prostration and death. The 1970–1973 panzootic caused by this type of virus, disease in some countries, such as Great Britain (Allan et al. 1978) and Northern Ireland (McFerran and McCracken 1988), was marked by severe respiratory signs, but in other countries these signs were absent. This type of VVND may cause oedema around the eyes and head. Green diarrhoea is frequently seen in birds that do not die early in infection; prior to death, muscular tremors, torticollis, paralysis of legs and wings and opisthotonos may be visible. Case fatality frequently reaches 100% in flocks of fully susceptible chickens.

The neurotropic velogenic form of ND (NVND) has been reported mainly in the United States. In chickens, it is marked by sudden onset of severe respiratory disease followed a day or two later by neurological signs. Egg production falls dramatically, but diarrhoea is usually absent. Morbidity may reach 100%. Case fatality in affected flocks is generally considerably lower, although mortality may be as high as 50% in adult birds and 90% in young chickens.

Mesogenic strains of ND virus usually cause respiratory disease in field infections. In adult birds, there may be a marked drop in egg production that can last for several weeks. Nervous signs are uncommon. Mortality in fowl is usually low, except in very young and susceptible birds, but may be considerably affected by exacerbating conditions.

Lentogenic viruses do not usually cause disease in adult birds. In young, fully susceptible birds, serious, often fatal respiratory disease problems occur following infection with the more pathogenic La Sota strains, especially in birds co-infected with other microorganisms. Vaccination or infection of broilers close to slaughter with these viruses can lead to colisepticemia or air sacculitis, with resulting condemnation.

The clinical signs produced by specific viruses in other hosts may differ widely from those seen in chickens. As the clinical manifestations of this disease are highly variable and no signs can be considered pathognomonic, absolute diagnosis is dependent upon the isolation and identification of the causative virus.

9.10 Differential Diagnosis

None of the clinical signs or lesions described above are specific for ND, which makes laboratory confirmation of a field diagnosis mandatory. The clinical signs and course of VVND may closely resemble those of a number of other avian diseases:

- Avian influenza (highly pathogenic type, HPAI)
- Fowl cholera
- Laryngotracheitis (acute form)
- Fowl pox (diphtheritic form)
- Ornithosis (psittacosis or chlamydophilosis) (psittacine birds and pigeons)
- Infectious bronchitis
- Pacheco's parrot disease (psittacine birds)
- Infections with avian paramyxovirus types 3 and 5 in some psittacine species
- Infectious bursal disease (Gumboro disease) (very virulent strains)
- Salmonellosis (pigeons)
- Other septicaemic infection (*Escherichia coli*, *Erysipelothrix rhusiopathiae*)
- Acute poisoning
- Management errors (deprivation of water, air, feed)

References

Aldous EW, Alexander DJ (2008) Newcastle disease in pheasants (*Phasianus colchius*): a review. Vet J 175(2):181-185

Alexander DJ (2000) Newcastle disease in ostriches (*Struthio camelus*) - A review. Avian Pathol 29(2):95-100

Alexander DJ, Senne DA (2008) Newcastle disease, other avian Paramyxoviruses and Pneumovirus infections. In: Saif Ym (ed) Diseases of poultry, 12th ed. Iowa State University Press, pp 75-100

Alexander DJ, Parsons G (1984) Avian paramyxovirus type 1 infections of racing pigeons: 2 pathogenicity experiments in pigeons and chickens. Vet Rec 114(19):466-469

Alexander DJ, Wilson GW, Thain JA, Lister SA (1984) Avian paramyxovirus type 1 infection of racing pigeons: 3 epizootiological considerations. Vet Rec 115(9):213-216

Allan WH, Lancaster JE, Toth B (1978) Newcastle disease vaccines – Their production and use. FAO Animal Production and Health Series No. 10. FAO: Rome, Italy

Beard CW, Hanson RP (1984) Newcastle Disease. In: "Diseases of Poultry" (Hofstad MS, Barnes HJ, Calnek BW, Reid WM, Yoder HW ed), 8th ed, Iowa State University Press: Ames, IA, pp 452-470

Bozorgmehri-Fard MH, Keyvanfar H (1979) Isolation of Newcastle disease virus from teals (*Anas crecca*) in Iran. J Wildl Dis 15(2):335-337

Erickson GA, Maré CJ, Gustafson GA et al (1977a) Interactions between viscerotropic velogenic Newcastle disease virus and pet birds of six species. I. Clinical and serologic responses, and viral excretion. Avian Dis 21(4):642-654

Erickson GA, Maré CJ, Gustafson GA et al (1977b) Interactions between viscerotropic velogenic Newcastle disease virus and pet birds of six species. II. Viral evolution through bird passage. Avian Dis 21(4):655-669

Estudillo J (1972) A Newcastle disease outbreak in captive exotic birds. Proc. 21st West. Poultry Dis. Conf., University of California, pp 70-73

Friend M, Franson JC (1999) Field manual of wildlife diseases. US Geological Survey, 001

Friend M, Trainer DO (1972) Experimental Newcastle disease studies in the Mallard. Avian Dis 16(4):700-713

Gerlach H (1994) Viruses. In: Avian Medicine: principles and application. Eds: Ritchie BW, Harrison GJ, Harrison LR, Lake Worth, Fl.: Wingers Publ. Inc.: pp 862-948

Gross WB (1961) The development of "air sac disease". Avian Dis 5:431-439

Huchzermeyer FW (1996) Newcastle Disease in ostriches in South Africa. Proc. Improving our understanding of rataties in a farming environment. Ed. Deeming DC, Manchester, England, pp 55-66

Kaleta EF, Baldauf C (1988) Newcastle disease in free-living and pet birds. In: Newcastle disease (Alexander DJ ed). Kluwer Academic Publishers, Dordrecht, Netherlands, pp 197-246

Kauker E, Siegert R (1957) Newcastlevirus – Infektion beim afrikanischen Strauß (*Struthio camelus*), Zwerggänsegeier (*Pseudogyps africanus Salvad.*) und Bunttukan (*Ramphastos dicolorus*). Monatshefte für Tierheilkunde, 9:64-68

Kim SJ, Spradbrow PB, MacKenzie M (1978) The isolation of lentogenic strains of Newcastle disease virus in Australia. Aust Vet J 54(4):183-187

Kingston DJ, Dharsana R, Chavez ER (1978) Isolation of a mesogenic Newcastle diseases virus from an acute disease in Indonesian ducks. Trop Anim Health Prod 10(3):161-164

McFerran JB, McCracken RM (1988) Newcastle disease. In "Newcastle Disease" (Alexander DJ ed), Kluwer Academic Publishers: Boston, MA, pp 161-183

Nakamura K, Ueda H, Tanimura T, Noguchi K (1994) Effect of mixed live vaccine (Newcastle disease and infectious bronchitis) and Mycoplasma gallisepticum on the chicken respiratory tract and on Escherichia coli infection. J Comp Path 111(1):33-42

Sallermann U (1973) Untersuchungen über die Newcastle-Krankheit beim Wellensittich (*Melopsittacus undulatus*) und die Möglichkeit einer Immunprophylaxe. Vet Diss, Gießen 1973

Samberg Y, Hadash DU, Perelman B, Meroz M (1989) Newcastle disease in ostriches (Struthio camelus): field case and experimental infection. Avian Pathology 18:221-226

Terregino C, Cattoli G, Vascellari M, Capua I (2004) Evaluation of efficacy of the vaccination against Newcastle disease in canary birds (*Serinus canarius*). Proceedings of the Fifty-Third Western Poultry Disease Conference March 7-9, 2004. Sacramento, California, pp 96-97

Verwoerd DJ (1995) Velogenic Newcastle disease epidemic in South Africa. Part II: ostriches, waterfowl, exotic bird collections and wild birds. S Af Vet Med 8:44-49

Conventional Diagnosis of Newcastle Disease Virus Infection

10

Calogero Terregino and Ilaria Capua

10.1 Virus Isolation in Embryonated Eggs

Virus is isolated following the protocol of the World Organisation for Animal Health (OIE) and European standards (OIE, 2008; CE, 92/66), and may be performed in embryonated eggs or in cell cultures. The procedure for virus isolation in embryonated eggs is identical to the one described for avian influenza (AI) in Chapter 7.

- A sample yielding a positive result in the rapid haemagglutinination (HA) test must undergo further tests, including titration of HA activity. This procedure enables the confirmation and quantification of HA activity, which are prerequisites to the identification of the HA agent by means of the haemagglutinatination inhibition (HI) test.

If a laboratory does not have the capacity to perform the HI test, the HA allantoic fluid should be sent to a National Reference Laboratory or to an International Reference Laboratory to confirm the diagnosis.

10.2 Cell Cultures

Newcastle disease (ND) virus strains can replicate in a variety of cell cultures of avian and non-avian origin, among those most widely used are: chicken embryo liver (CEL) cells, chicken embryo fibroblasts (CEF), African green monkey kidney (Vero) cells and chicken-embryo-related (CER) cells. Viral growth is usually accompanied by damage of the monolayer, known as the cytopathic effect (CPE). Due to the properties of ND Virus, viral replication results in disruption of the monolayer by the formation of syncytia, which are large multinucleate cells originating from the fusion of several cells. Plaque formation in chick embryo cells is restricted to velogenic and mesogenic viruses unless Mg^{2+} ions and diethylaminoethyl dextran (DEAE) or trypsin are added to the culture medium.

Some strains of pigeon paramyxovirus type 1 (PPMV-1) are difficult to isolate in embryonated eggs but grow well in CEL cells; thus, virus isolation from samples suspected of being infected with PPMV-1 should be attempted using both substrates.

Since the viral titre that can be obtained in cell culture is generally very low, the isolates should be passaged into embryonated eggs prior to characterisation of the virus.

For a detailed description of the haemagglutination test in Petri dishes (rapid HA test), see Chapter 7 page 73.

10.3 Characterisation of Newcastle Disease Virus by Serological Methods

Newcastle disease virus isolates are a frequent finding in samples submitted to avian virology laboratories. These isolates consist of virulent viruses causing disease in the field and nonvirulent viruses used as live vaccines or naturally present in birds. It is therefore essential that diagnostic laboratories are equipped to identify ND isolates and to differentiate them from other HA agents. Allantoic fluids with HA activity should be further tested by means of the HI test. Viral HA activity of viruses grown in fowl's eggs may be due to any of the nine avian paramyxovirus serotypes or to any of the 16 influenza type A haemagglutinin subtypes that are known to infect birds. Viral identity can be ascertained by performing an HI test using specific polyclonal antisera. In case of identity between the isolate and the antiserum, the HA activity of the virus is inhibited. However, care must be taken in interpreting the results as cross-reactions

I. Capua, D.J. Alexander (eds.) *Avian Influenza and Newcastle Disease*,

may occur between viruses belonging to different serotypes. For example, ND virus (APMV-1) has been reported to show some degree of cross-reactivity in HI tests with several of the other avian paramyxovirus serotypes, especially APMV-3 psittacine isolates, using polyclonal antisera. The risk of misdiagnosing an isolate can be eliminated using control sera and a panel of reference antigens.

Confirmatory diagnosis of ND is based on viral characterisation with the HI test, as described in Chapter 7, using reference avian paramyxovirus subtype antisera.

10.4 Serological Tests To Be Used for the Detection OF Specific Antibodies Against Newcastle Disease Virus

10.4.1 Introduction

The detection of specific antibodies to ND virus is a routine activity of avian virology laboratories. Generally speaking, serological assays are carried out to evaluate the immune response following vaccine administration or to detect seroconversion following natural infection. The latter may be ascertained following the detection of antibodies in naïve (unvaccinated) birds or by an increase in the antibody titre (at least four-fold) in vaccinated birds.

The most commonly used quantitative assay is the HI test, following the same procedure described for AI except that the reference antigen contains APMV-1 viruses. The HI test is also considered the gold standard for the validation of any other serological assay.

10.4.2 Haemagglutination Inhibition (HI) Test

For a detailed description, see Chapter 7 (HI test for AI).

10.4.3 Evaluation of Pathogenicity Using Conventional Methods

10.4.3.1 Classification Based on Clinical Signs

Five groups of APMV-1 strains have been distinguished according to the clinical signs observed in experimentally infected specific-pathogen free (SPF) chickens (Beard and Hanson 1984). Virulent strains causing high mortality with haemorrhagic lesions in the gastrointestinal tract are designated as viscerotropic velogenic (VVND), while those causing high mortality with respiratory and nervous signs are neurotropic velogenic (NVND). Mesogenic strains are characterised by causing respiratory and nervous signs in infected animals but usually with low mortality. Lentogenic strains cause typically mild or unapparent infection of the respiratory tract while asymptomatic enteric strains produce an unapparent intestinal infection.

10.4.3.2 Mean Death Time and Plaque Formation

Estimation of virulence based on the varying lethal effect of viruses on chicken embryos was originally described by Hanson and Brandly (1955). The mean death time (MDT) test is based on the experience that virulent viruses kill embryos quicker than those with lower virulence. Velogenic strains kill embryos in less than 60 h, mesogenic strains in 60–90 h and lentogenic strains in > 90 h (Alexander 1988). Currently, the MDT assay is not considered a reliable test for the characterisation of ND viral isolates associated with outbreaks, and it is not included in Directive 92/66/EEC.

The capability of APMV-1 to form plaques can be investigated in primary CEF cells or in cultured lung or kidney cells. Plaque size, morphology and colour are related to the degree of virulence. Lentogenic strains require the addition of DEAE and magnesium ions or trypsin for plaque formation.

10.4.3.3 Intracerebral Pathogenicity Indices

The most sensitive and widely used test for measuring virulence is the intracerebral pathogenicity index (ICPI) in day-old chickens.

Procedure

1. Dilute infective freshly harvested allantoic fluid (HA titre > 2^4) 1: 10 in sterile isotonic saline (antibiotics must not be used).
2. Inject 0.05 ml of the diluted virus is intracerebrally into each of ten 1-day-old chicks (i.e. 24–40 h after hatching). The chicks should be hatched from eggs obtained from an SPF flock.

3. The birds are examined at intervals of 24 h for 8 days.
4. At each observation, each bird is scored: 0 = normal, 1 = sick, 2 = dead.
5. The index is calculated as shown in the following example. A simple method for recording results and calculating indices is shown in Table 10.1:

Table 10.1 Determination of the intracerebral pathogenicity index (ICPI)

Clinical signs	Days after inoculation Number of chickens with specific signs								Total score
	1	2	3	4	5	6	7	8	
Normal	10	4	2	0	0	0	0	0	16 × 0 = 0
Sick	0	6	5	6	2	0	0	0	19 × 1 = 19
Dead	0	0	3	4	8	10	10	10	45 × 2 = 90
									TOTAL = 109/80
									ICPI = 1.36

10 birds observed for 80 days = 80 observations
Index = mean score per bird per observation = 109/80 = 1.36
Any APMV-I yielding a value of 0.7 or greater in an ICPI test is considered to be a virulent ND virus.
The ICPI is the mean score per bird per observation over the 8-day period. The most virulent isolates have an ICPI close to 2.0, lentogenic and asymptomatic enteric viruses have values of 0.0–0.6.

References

Alexander DJ (1988) Newcastle disease virus - An avian paramyxovirus. In: Newcastle Disease, Alexander DJ ed., Kluwer Academic Publishers: Boston, MA, pp 11-22

Beard CW, Hanson RP (1984) Newcastle Disease. In: Diseases of Poultry, Hofstad MS, Barnes HJ, Calnek BW et al ed., 8th ed., Iowa State University Press: Ames, IA, pp 452-470

CEC (1992) Council Directive 92/66/EEC of 14 July 1992 introducing Community measures for the control of Newcastle disease. Official Journal of the European Communities L 260

Hanson RP, and Brandly CA (1955) Identification of vaccine strains of Newcastle disease virus. Science 122(3160):156-157

OIE (2008) Newcaslte Disease. World Organisation for Animal Health Manual of Diagnostich tests vaccines for terrestrial animals, 6th ed. Chapter 2.3.14. OIE Paris, pp 576-589

Molecular Diagnosis of Newcastle Disease Virus

11

Giovanni Cattoli and Isabella Monne

11.1 Introduction

Several protocols for the detection of avian paramyxovirus type 1 (APMV-1) by retrotranscription polymerase chain reaction (RT-PCR) have been published in the last decade. A technical review was presented a few years ago (Aldous et al, 2001) in which conventional (end-point) RT-PCR protocols were described. More recently, real time PCR (rPCR) protocols based on the use of hydrolysis probes, SybrGreen or LUX primers have been published (Wise et al. 2004; Pham et al. 2005; Antal et al. 2007). In the following, protocols used by the IZSVe are presented.

For details concerning the type of samples to be tested, sample preparation and RNA extraction, refer to the previous section on avian influenza.

11.2 Detection and Typing of APMV-1 by End-Point RT-PCR and Restriction Endonuclease Analysis

11.2.1 Protocol 1

This one-step RT-PCR protocol has been adapted at the IZSVe and is based on the primer set described by Creelan et al. (2002). In this protocol, amplification of APMV-1-specific nucleic acid fragments followed by enzymatic digestion (restriction endonuclease analysis, REA) using *Bgl*I was carried out to detect and type strains according to their virulence. In the original paper, compared to virus isolation, the relative sensitivity of this protocol for clinical specimens was 73.44% on a sample basis, rising to up to 91.30% on a case basis (Creelan et al. 2002).

Due to the small amplicon size, it is recommended that the RT-PCR results be visualised on silver-stained SDS-polyacrylamide gels or 2–3% agarose gels.

cont.

I. Capua, D.J. Alexander (eds.) *Avian Influenza and Newcastle Disease,*

One-step RT-PCR (AB 9700 thermal cycler) (Super script One step RT-PCR with Platinum Taq, Invitrogen # 10928-042 or –034)

Target: F gene
Sample: 5 µl RNA in a total reagent volume of 45 µl

Primers

Forward NDV-F 4829: 5'-GGTGAGTCTATCCGGARGATACAAG-3'
Reverse NDV-R 5031: 5'-TCATTGGTTGCRGCAATGCTCT-3'

Procedure

Reagent[a]	Final concentration	1 × Reaction (µl)
RNase-free water	/	16
PCR Buffer 2 × ($MgSO_4$ 2.4 mM; dNTPs 1.6 mM)	1X	25
Primer NDVF 4829	0.2 µM	1.0
Primer NDVR 5031	0.2 µM	1.0
Enzyme mix (RT and Taq)	/	1
Total reagent volume		45
Vortex the mixture for a few seconds.		
Aliquot 45 µl into 0.2-ml PCR tubes.		
Add RNA		5
Final volume		50

[a]Concentration of stock solution

Cycling Conditions

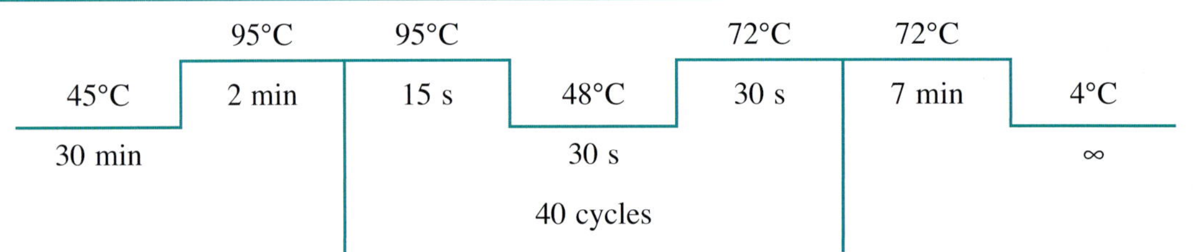

Detection: Silver-stained SDS-polyacrylamide 7% gel or agarose gel 2%.
Expected amplified fragment: 202 bp.

Restriction endonuclease analysis (**REA**)
Purify the cDNA product using a commercial kit (e.g., High Pure PCR product Purification kit, Roche) following manufacturer's instructions.
Digest the PCR amplification product (5 µl in a final volume of 50 µl) with 10 U of *Bgl*I (Roche, Germany; # 621641) and the appropriate buffer (Buffer H, Roche) for 2.30 h at 37°C.
Visualize the results after gel electrophoresis.
Expected results
Lentogenic APMV-1 restriction pattern: 2 bands of approximately 135 and 67 bp
Meso- and velogenic APMV-1 restriction pattern: 1 band of approximately 202 bp (i.e. no digestion).

11.3 Detection and Typing of APMV-1 by End-Point RT-PCR

The following protocols (2 and 3) are used at the IZSVe for the genotyping and pathotyping of APMV-1 isolates. They couple RT-PCR amplification with genetic sequencing and can also be used to confirm the presence of APMV-1 in allantoic fluid. The direct use of these protocols on diagnostic specimens has never been fully evaluated. Protocols 2 and 3 include different primer sets. Together, these protocols are capable to amplify representative strains of all the genetic lineages of APMV-1 described so far.

11.3.1 Protocol 2

NOH One-step RT-PCR (AB 9700 thermal cycler) (OneStep RT-PCR kit ; Qiagen) for the detection of APMV-1 RNA in the allantoic fluid of embryonated fowl eggs

Primers

Forward NOH-For 5' TACACCTCATCCCAGACAGG 3'
Reverse NOH-Rev 5' AGTCGGAGGATGTTGGCAGC 3'

Procedure

Reagent[a]	Final concentration	1 × Reaction (µl)
RNase-free water	/	26.8
PCR buffer 5×	1X	10
dNTP mix 10 mM	0.2mM	1
Primer NOH-For 100 µM	1 µM	0.5
Primer NOH-Rev 100 µM	1 µM	0.5
RNase Inhibitor 40 U/µl	8U	0.2
One-Step RT-PCR enzyme mix		1
Total reagent volume		40
RNA		10
Final volume		50

[a]Concentration of stock solution

Cycling Conditions

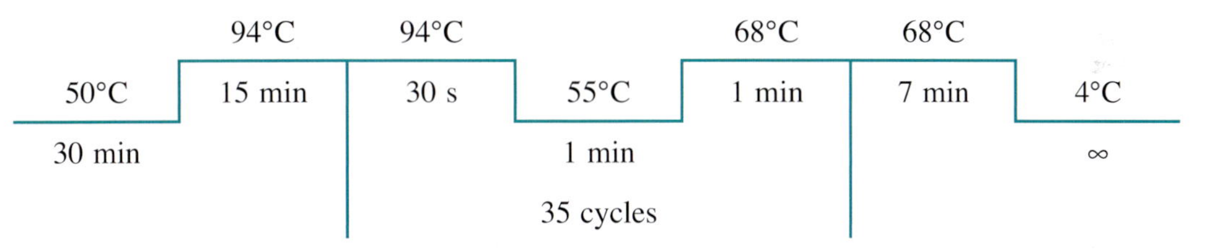

Detection: Silver-stained SDS-polyacrylamide 7% or agarose gel 2%
Expected amplified fragment: Approximately 300 bp

11.3.2 Protocol 3

H One-step RT-PCR (AB 9700 thermal cycler) (GeneAmp Gold RNA PCR core kit, Applied Biosystems # 4308207) for the detection of APMV-1 RNA in the allantoic fluid of embryonated fowl eggs

Primers

Forward H-For 5' ATGCCCAAAGACAAAGAGCAA 3'
Reverse H-Rev 5' TACTGCTGTCGCTACACCTAA 3'
Pre-RT: 10 μl of RNA at 60°C for 10 min

Procedure

Reagent[a]	Final concentration	1 × Reaction (μl)
RNase-free water	/	17.9
PCR buffer 5×	1×	10
$MgCl_2$ 25 mM	1.25 mM	2.5
dNTPs mix 10 mM	0.8 mM	4
Primer NDV-H-for 12.5 μM	0.25 μM	1
Primer NDV-H-rev 12.5 μM	0.25 μM	1
DTT 100 mM	5 mM	2.5
RT enzyme 50 U/μl	15 U/50 μl	0.3
Rnase inhibitor 20 U/μl	10 U/50 μl	0.5
AmpliTaq Gold 5 U/μl	1.5 U/50 μl	0.3
Total reagent volume		40
RNA		10
Final volume		50

[a]Concentration of stock solution

Cycling Conditions

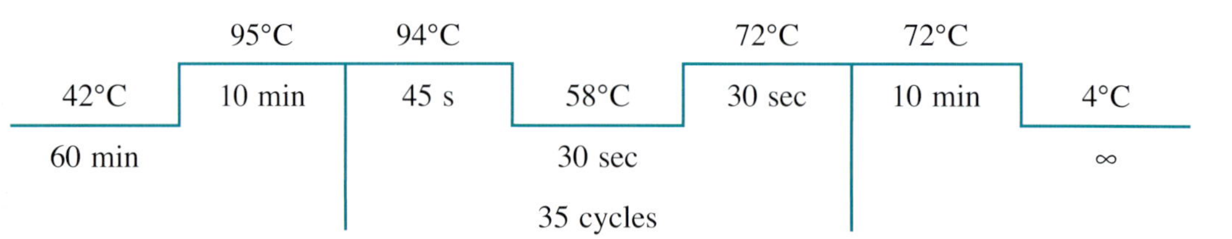

Detection: Silver-stained SDS-polyacrylamide 7% or agarose gel 2%
Expected amplified fragment: approximately 300 bp

11.4. Detection of APMV-1 by Qualitative One-step Real-time RT-PCR

11.4.1 Protocol 4

One-step real time RT-PCR (QuantiTect Multiplex RT-PCR kit 2X, # 204643) for the detection of APMV-1 RNA in clinical specimens

Target: Matrix gene
Sample: RNA 5 μl in a reaction volume of 20 μl

Primers and probe *(Wise et al. 2004)*
APMV1 F 5'-AGT GAT GTG CTC GGA CCT TC-3'
APMV1 R 5'-CCT GAG GAG AGG CAT TTG CTA-3'
Probe APMV1 5'-[FAM] TTC TCT AGC AGT GGG ACA GCC TGC [TAMRA]

Procedure

Reagent[a]	Final concentration	1 × Reaction (μl)
RNase-free water	/	4.9
Primer APMV 1F 10 μM	400 nM	1
Primer APMV 1R 10 μM	400 nM	1
2× RT-PCR master mix	1×	12.5
Probe FAM APMV1 10 μM	160 nM	0.4
Enzyme mix		0.2
Total reagent volume		20
RNA		5
Final volume		25

[a]Concentration of stock solution

Cycling Conditions

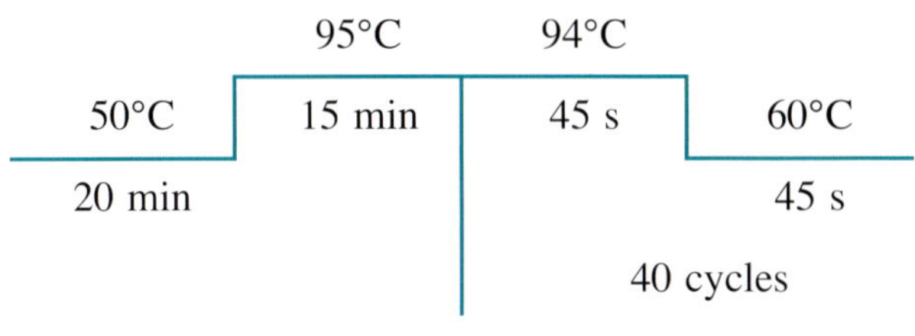

References

Antal M, Farkas T, Germán P et al (2007) Real-time reverse transcription-polymerase chain reaction detection of Newcastle disease virus using light upon extension fluorogenic primers. J Vet Diagn Invest 19(4):400-404

Creelan JL, Graham DA, McCullough SJ (2002) Detection and differentiation of pathogenicity of avian paramyxovirus serotype 1 from field cases using one-step reverse transcriptase-polymerase chain reaction. Avian Pathol 31(5):493-499

Aldous EW, Alexander DJ (2001) Detection and differentiation of Newcastle disease virus (avian paramyxovirus type 1) Avian Pathol 30:117-128

Pham HM, Konnai S, Usui, T et al (2005) Rapid detection and differentiation of Newcastle disease virus by real-time PCR with melting-curve analysis. Arch Virol 150(12):2429-2438

Wise MG, Suarez DL, Seal BS et al (2004) Development of a real-time reverse-transcription PCR for detection of Newcastle disease virus RNA in clinical samples. J Clin Microbiol 42(1):329-338

12 General Rules for Decontamination Following an Outbreak of Avian Influenza or Newcastle Disease

Maria Serena Beato and Paola De Benedictis

12.1 Introduction

Rapid application of strict biosecurity measures is the first step to prevent and control the introduction of avian influenza (AI) or Newcastle disease (ND) viruses. Biosecurity comprises two elements: bio-exclusion and bio-containment. Bio-exclusion includes all measures aimed at excluding infectious agents from uninfected premises. It requires the prevention of direct and indirect contact of infected animals or contaminated inanimate carriers (*fomites*) with poultry. Bio-containment includes all measures aimed at maintaining the infection within the premises from where the diagnosis was first obtained. Decontamination of the infected farm is one of the actions that must be adopted during the bio-containment process (EFSA 2005). Secondary spread of AI and ND is achieved mainly through human-related activities, such as the movement of staff, vehicles, equipment and other fomites. Further outbreaks may occur following restocking of birds in establishments that have not been adequately sanitised. It therefore follows that if decontamination of premises, footwear and clothing, vehicles, crates, farm equipment and other materials is not carried out properly, infection will persist in the avian population. Thus, the concurrent damage to the poultry industry and, in many instances, the public health threat will not be removed. For this reason, cleaning and disinfecting must be considered as an essential part of AI and ND control programmes. Decontamination is the combination of physical and chemical processes that kill or remove pathogenic microorganisms and is of crucial importance for disease eradication. Decontamination involves close cooperation between property owners and the personnel involved in the procedures. Natural processes, such as time, dehydration, warm temperature and sunlight, favour decontamination. Since most disinfectants have reduced effectiveness in the presence of fat and organic matter, preliminary cleaning is needed invariably before any disinfection, in order to achieve effective chemical decontamination.

12.2 Choice of Disinfectant

When a disinfection programme is to be implemented, many factors must be taken into account in order to achieve the goal of decontaminating the infected area and, in the meantime, limiting spread of infection to uninfected farms. Knowledge of the characteristics of infectious agents plays a key role in the selection of the disinfectant. AI and ND viruses are multiplied and released at a high concentration by their host species and are able to persist in the environment (Beard and Hanson 1984; De Benedictis et al. 2007). Nevertheless, their resistance to common disinfectants is relatively low, according to the Noll and Youngner classification (1959). Both AI and ND viruses are medium-sized, single-stranded (ss) RNA, enveloped viruses. They are classified as Category A viruses (Noll and Youngner 1959).

However, when planning a disinfection programme, other factors also need to be taken into account, namely the properties of disinfectants, external factors that can influence the activity of the disinfectants and the characteristics of the premises to be decontaminated. Choice of the appropriate disinfectant relies on its proven efficacy against AI and ND viruses. The assessment should also include sustainability under certain circumstances (i.e. pH and temperature for optimal activity, stability in the presence of organic material or in hard water, con-

I. Capua, D.J. Alexander (eds.) *Avian Influenza and Newcastle Disease*,

tact time and safety for personnel). Products should be used at a concentration of proven efficacy against the selected agent. Most disinfectants require a minimum contact time, usually not less than 30 min, depending on environmental conditions. The toxicity of some chemical agents limits the range of choice. The use of a compound with a high toxicity should be avoided although compounds with mild toxicity may be used under controlled circumstances. It is of primary importance that the staff members are trained in disinfecting procedures in order to achieve optimal results and avoid adverse consequences to operators, equipment and the environment. Corrosiveness can be limited by diluting the compound whilst taking into account the minimum concentration necessary to maintain viricidal activity. In many instances, the cost and availability, at a local level, of the disinfectant product are the main limiting factors.

The best inactivating agents for Category A viruses, including AI and ND viruses, are considered to be detergents, alkalis, oxidising agents, and aldehydes (Ausvetplan 2007). The use of soaps and detergents is recommended only for preparatory clean-up procedures before proper decontamination. Alkalis are ideal decontaminating chemicals for animal housing, yards, drains, effluent waste pits and sewage collection areas. Sodium hydroxide, caustic soda and sodium carbonate washing soda are readily available, cost-effective and have a saponifying action on fats and organic matter. In contrast, oxidising agents are not recommended for decontamination procedures. The effectiveness of household bleach (sodium hypochlorite) and hypochlorite powder decreases markedly in the presence of organic matter; these compounds are also not stable chemically and decompose rapidly at temperatures above 15°C. Commercial products are highly effective, although expensive, and should be used according to the manufacturer's instructions. Within the aldehyde family, gluteraldehyde is efficacious, stable and partially active in the presence of organic matter, although it is a mild corrosive for metals. Despite these positive characteristics, the cost of gluteraldehyde for large-scale decontamination is high.

Gaseous formaldehyde is still used for the decontamination of air spaces. However, many parameters have to be assessed to achieve a complete and effective decontamination, including gas concentration, temperature, humidity, contact time and evenness of distribution (see Appendix A for practicalities of formaldehyde gas use).

Other external factors that should be considered to achieve optimal efficacy of the decontamination process are the calcium concentration in the water used to prepare the disinfectant solution and the environmental temperature, both of which influence the efficacy of the disinfectant. In general, disinfectants are most efficacious at high temperatures, reaching optimum efficacy above 20°C (e.g. the optimum range for formaldehyde activity is 24–38°C) (Samberg and Meroz 1995). Some products effective against AI have been tested in combination with antifreeze compounds and shown to retain their activity (Davison et al. 1999). During the winter or in case of low environmental temperatures, the efficacy of certain disinfectants may be reduced. It is essential that the product to be used under these conditions is still efficacious at low temperatures or retains its activity when combined with an antifreeze product.

A summary of disinfectants to be used in decontamination procedures is summarised in Tables 12.1 and 12.2.

Table 12.1 Disinfectant/chemical selections and procedures for avian influenza andNewcastle disease (modified from Ausvetplan 2007)

Item to be disinfected	Disinfectant/chemical/procedure
Live bird	Kill humanely
Carcases	Bury, burn or render
Animal housing/equipment	Soaps and detergents, oxidising agents, alkalis
Environs	N/A
Humans	Soaps and detergents, citric acid
Water	
– Tanks	Drain to pasture where possible
– Dams	Drain to pasture if practicable, otherwise N/A
Electrical equipment	Formaldehyde gas
Feed	Bury, burn
Effluent, manure	Bury or burn, alkalis and acids
Human housing	Soaps and detergents, oxidising agents
Machinery, vehicles	Soaps and detergents, alkalis
Clothing	Soaps and detergents, oxidising agents, alkalis
Aircraft	Soaps and detergents, oxidising agents

Table 12.2 Chemical products available for disinfecting procedures and principal recommendations of use (modified from De Benedictis et al. 2007). Recommended products are highlighted in bold type

Chemical product	Recommended concentration	Method of action	Recommended contact time	Recommended use	Limitations	Other information
Soaps and detergents		Surfactant propriety against lipid components	10 min	During cleaning		Used also with disinfectants
Alkalis		Protein denaturation			Activity increases at high temperature; not efficacious at room temperature	
Sodium hydroxide (caustic soda)	2–5% for clothes 10% at 60° C for floors		10 min	Floors and clothes	Do not use in the presence of aluminium and derived alloys	Do not allow contact with organic tissues
Sodium carbonate	10%		30 min	In the presence of high concentrations of organic material	Thermolabile; light-sensitive	
Calcium hydroxide	3%			Walls, floors		
Acids		Inhibition of enzymatic reactions; denaturing of proteins and nucleic acids				
Hydrochloric acid (inorganic acid)	2–5%		10 min	Floors	Corrosive; do not use for disinfecting metals	
Citric acid (organic acid)	0.2%		30 min	Clothing and body		
Chlorine compounds		Protein denaturation and oxidising			Corrosive; inhibited by organic materials and by basic pH	Low cost and non toxic
Calcium hypochlorite	2–3%		10–30 min	Floors, clothes		
Sodium hypochlorite (household bleach)	2–3%		10–30 min	Equipment		
Oxidising agents		Denaturing activity on lipids and DNA			Decreasing efficacy in the presence of organic compounds; corrosive	
Hydrogen peroxide	3–6%					Rinse after use
Aldehydes		Alkylation of amino and sulphydryl groups of protein and of nitrogen of purine bases			Decreasing efficacy in the presence of organic compounds; corrosive	

(*continued*)

Table 12.2 *(continued)*

Chemical product	Recommended concentration	Method of action	Recommended contact time	Recommended use	Limitations	Other information
Formalin	8%		10-30 min		Toxic gas; unstable	Efficacious in the presence of propylene glycol
Glutaraldehyde	1–2%		10–30 min		pH 7.5–8.5: mildly corrosive for metals, not for use on plastic and rubber	Irritating for eyes, nose and throat; a rinse after use
Formaldehyde	40%		15–24 h			
Phenol compounds		Inactivation of enzymatic system and loss of metabolites through cellular membrane			Irritating due to their residual activity: rinse after use	Efficacious in the presence of organic matter; low cost
Cresolic acid	2%			Floors		High cost
Synthetic phenols	2%		10 min	Floors		
Phenol crystal	0.4–0.2%		12–18 h			
Quaternary ammonium compounds		Activity with –NH4+ groups		Personal use	Do not use with hard water, e.g. > 32 F°; efficacious in the presence of antifreeze compounds	Accurate cleaning of surfaces is recommended before use
Alcohols		Protein denaturising in the presence of H_2O		Clothes and equipment	Do not use for plastic and rubber	Flammable, evaporable
Ethanol	70%		5–15 min	In association with other compounds in hand-wash disinfectants		Used also in association with other molecules or as a thinner in disinfectant solutions

12.3 Decontamination Procedures

Effective property decontamination will be achieved as a result of appropriate assessment of the contaminated areas and extensive knowledge of the characteristics of the infectious agent. Further requirements are the availability of adequate equipment, disinfectants and personnel to undertake the tasks. As a preliminary good practice, all exhaust fans must be turned off in the case of an outbreak occurring in an intensive poultry farm. This is of primary importance to avoid uncontrolled dispersion of the agent by aerosol (Ausvetplan 2007).

A decontamination strategy consists of:

- Property assessment
- Preliminary disinfection
- Initial clean-up
- Full disinfection followed by inspections

This includes disinfection of personnel leaving the contaminated areas as well as infected areas and machineries. Special attention must be paid to areas at high risk of contamination, such as animal waste effluents and animal feed. The latter must be destroyed.

Both cleaning and disinfection procedures of the infected premises must be performed systematically from back to front and from top to bottom of the farm. The roof-wall-floor method should be adopted in each building and the building should be cordoned off with marking tape when disinfection is concluded in order to avoid re-contamination of a decontaminated area (De Benedictis et al. 2007; Ausvetplan 2007).

Preliminary disinfection should be undertaken immediately after confirmation of the disease, as it will reduce the amount and distribution of infectious agent during culling and disposal. This preliminary disinfection should be performed in any contaminated area, with particular attention paid to culling and at disposal sites. In particular, the culling site should be continuously disinfected at every break during the day (Ausvetplan 2007).

After slaughter and disposal, a clean-up process should be undertaken to remove all manure, dirt, detritus and contaminated items that cannot be disinfected, e.g. insulation material, wood, contaminated feedstuff and litter. The use of water and disinfectant should be avoided at this stage, to reduce both the volume and the weight of the material to be disposed of (dry cleaning). After disposal, all surfaces should be scratched, scraped, and then sprayed with low-pressure water and detergent to remove any visible contamination. Earthen floors should be broken and soaked in disinfectant (wet cleaning). The viricidal activity of the majority of disinfectants is inhibited partially or totally by interaction with organic material. For this reason, thorough clean-up should be considered an essential initial step for an efficacious disinfection.

During full disinfection procedures, the goal should be the inactivation of all infectious particles. Portable equipment (platforms, feeding-trough, egg rollers, egg conveyors, egg collectors) must be cleaned and then disinfected indoors to prevent any contact with uninfected livestock. Water pipes must be flushed with high-pressure water and then all parts of the pipes filled with a water solution of disinfectant for at least 48 h. Pipes have to be rinsed with a further water jet. Water pipes that can be dismantled must be cleaned individually with cleaning solutions collected directly into containment vessels. To decontaminate iron fittings, the use of high temperature and, if safety considerations allow, the application of a flame are recommended.

A thorough inspection provides an assessment of the efficacy of decontamination. Important aspects to be checked are:

- Complete disposal of all contaminated woodwork not suitable for cleaning and disinfection
- No organic material is left behind fixtures and fittings
- No encrustation on any exposed surface is observable
- All contaminated feedstuff has been destroyed
- All grossly contaminated sites (culling and disposal) have been cleaned effectively and disinfected
- All fluid that has been disinfected has been released into drains or septic tanks
- The conditions of quarantine, especially at exit/entry points, and warning notices are maintained.

The second disinfection is a repeat of the first and can be started approximately 14 days after the completion of the first disinfection. Final inspection is carried out in the same way as the first inspection. The workforce is to be withdrawn from the premises only if the inspection yields positive results and there are no doubts on its effectiveness and completeness. If there is any degree of uncertainty, the procedure must be repeated. The efficacy of the disinfection process may be tested by introducing sentinel animals or by collecting environmental swabs for virus isolation attempts.

12.4 Personal Decontamination

During an AI or ND outbreak, people may spread the virus by acting as mechanical carriers. For this reason, it is necessary that all staff members taking part in the decontamination procedures change clothing, use disposable shoes and overalls before entering the farm and shower when they leave the infected premises. Heavy personal contamination occurs inevitably whilst working on infected/contact premises, particularly during physical inspection of living animals, at culling and carcase disposal sites, and when removing manure, bedding and detritus. A personal decontamination site (PDS) must be arranged near the exit point of an infected premise (IP), and moved into the IP when necessary. The PDS should be placed at the limit of the total area defined as infected in order to avoid secondary contamination of people leaving the PDS. The PDS should be easily disinfected and have an impervious surface; alternatively, the floor area may be covered with a large plastic ground cover. Treatment (usually spraying) with an efficacious disinfectant should be undertaken before any procedures are started. Clean water and good drainage are crucial to avoid recontamination of clean areas. If adequate drainage is not available, a pit may be used as an alternative to ensure that no effluent escapes beyond the decontamination site. Personal decontamination procedures must be followed strictly by all personnel leaving the IP. On arrival at the PDS, warm soapy water should be available for washing the hair, face and skin. The pH of the water solution can be varied to enhance its antiviral action, with the addition of sodium carbonate or citric acid. Heavy-gauge plastic garbage bags should be used for the storage of all contaminated items. Plastic bags are easily disinfected by spraying their external surfaces; this procedure avoids further contamination of personnel leaving the IP to burn and bury waste or to clean and disinfect non-disposable items. When available, the use of disposable overalls must be favoured over other clothes. Plastic overalls should first be washed with a low-pressure pump to remove gross material. Particular care must be taken to clean the back, under the collar, the zipper and inside the pockets. Cotton overalls and sprayed plastic overalls are removed and placed in disinfectant. Underwear also should be placed in disinfectant, especially if cotton overalls are used. In this case, washing of the entire body is also necessary. Boots must be scrubbed, particularly the soles. Personnel leaving the PDS should walk across the areas, treat the boots again and finally change them for street shoes. Personnel are recommended to continue a second phase of cleaning at home. It is compulsory that they do not have direct or indirect contact with other susceptible animals, premises and poultry farms for a minimum of 3 days. Disinfected overalls must be placed in a plastic bag, the outside of the bag disinfected and then placed at the outer limit of the area for removal. The disposed items should be autoclaved or treated in a hospital laundry.

Visitors on properties where AI or ND is suspected should also be considered as contaminated. They should remain preferably in the suspected area until outbreak confirmation and the start of decontamination procedures. Otherwise, common household disinfectants should be used to minimise the risk of disease transmission. In this case, the following information should be recorded and advice given:

- Name and address of the people concerned
- Assessment of the degree of exposure and contact with the suspected disease agent
- Advise a change of clothing if possible
- Recommend putting the clothes suspected of contamination in a plastic bag for appropriate treatment
- Efficacious domestic chemicals, in default of approved disinfectants, are:
  - Domestic washing soda (10 parts in 100 parts hot water)
  - Soap and hot water for scrubbing
  - Household concentrated chlorine bleach (1 part in 3 parts of water, corresponding to 2–3% of available chlorine). This is not recommended for decontamination of the skin.

12.5 Vehicle and Car Decontamination

All vehicles that enter the IP, and their drivers, carry a disease dissemination risk. No vehicle may leave the IP before its decontamination. Additionally, all vehicles that have been in contact with the disease agents before the outbreak must be traced to avoid secondary and uncontrolled spread of the infection. A carwash facility is ideal for the decontamination of vehicles. It has the advantage of allowing the undercarriage of the vehicles to be very easily washed, thus cleansing the most contaminated part of the vehicle.

Any rubber floor mats should be removed and scrubbed with disinfectant. The dashboard, steering wheel, handbrake, gear stick and seats should be wiped with appropriate disinfectant. The contents of the boot must be removed and both the contents and the interior of the boot wiped with disinfectant. The wheels, wheel arches and undercarriage of the car should be sprayed with disinfectant. Cleaning using disinfectant/soap and water with brushing to dislodge encrusted dirt and organic matter is preferable to washing with strong water streams.

All solid debris should be removed from the vehicle. Livestock vehicles are then soaked in disinfectant using a detergent, and scrubbed down to bare metal or wood. The outside dual wheels and spare wheels must be removed to ensure adequate decontamination of wheel hubs and to inspect the spare wheel hangers. All animal faecal matter and bedding must be removed. All organic material must be considered as contaminated and then disinfected and burnt or buried. All fixtures and fittings must be dismantled to ensure that infected material has been removed. All surfaces must be cleaned and then disinfected. The wheels, wheel arches, bodywork and undercarriage must be cleaned of detritus and disinfected. The driver's cabin and the sleeping compartments also need to be cleaned and disinfected.

12.6 Disposal of Carcases

This section of the chapter briefly describes and then summarises the main methods for the disposal of animal carcases. Interesting and specific literature based on field experience gained during the management of outbreaks such as the 1984 AI outbreak in Virginia (US) and the 2001 FMD outbreak in the UK is available (Berglez 2003; UK Environment Agency 2001). Readers should take into account that all such information has to be applied flexibly, as carcase disposal is a part of an emergency management plan based on the specific options for disposal. Decision-makers should be knowledgeable about the various disposal technologies, understand their principles of operation, and be aware of the equipment needed, costs, environmental impact and logistic details for each technology. This expertise is best achieved by the formation of an *ad hoc* team of experts.

The primary aim in the disposal of carcases and animal products is to limit disease spread. In view of this, the disposal of carcases should be considered as an essential component of animal disease control and eradication programmes. To maintain biosecurity standards and decrease the risk of disease spread, it is necessary to know the epidemiology of the infectious agent as this will affect the choice of disposal methods. Regardless of the chosen method, rapid disposal and the classification of wastes according to their potential infectivity are of primary importance.

The methods used to dispose of animals and animal products and the selection of disposal sites must be based on the following principles. Before a particular plan of action is decided upon, a decision-making process incorporating these principles should be undertaken (Ausvetplan 2007):

- Prevention of disease spread
- Speed
- Cost effectiveness
- Local legislative requirements
- Community and operator safety
- Local environmental conditions and resource availability.

Selection of an expert team can be evaluated to analyse the field situation and to guide a decision-making process that yields recommendations allowing application of the best practicable solution at a local level.

After the carcases have been disposed of, long-term factors must be considered and planned, such as maintenance, monitoring and the rehabilitation of disposal sites. Carcases may be buried, incinerated, composted or rendered.

12.6.1 Burial

There are three burial techniques: (1) trench burial, (2) landfill and (3) mass burial sites.

12.6.1.1 Trench Burial

This approach involves excavating a trench, placing carcases in it and then using the excavated material to cover them. Since little expertise is required, this method is used widely. It is relatively inexpensive as most of the equipment necessary is readily available. Trench burial is generally adopted on-farm or on-site for daily mortalities and is probably more discrete than other methods such as open burning. Cost estimates of use of on-site trench burial may differ considerably when carried out in an emergency situation. In choosing this method of disposal, it is necessary to determine the suitability of a site for burial. Soil properties, topography, hydrological properties, proximity to water bodies, public areas, roadways, municipalities and property lines as well as accessibility affect the choice and thus the use of a site for burial. The disadvantages of this method include potential environmental contamination, especially of water. Regions where the water table is deep and the soil relatively impermeable are suitable for trench burial disposal. This method has been identified as a means of placing carcsses "out of site out of mind" (NABC 2004) while they decompose, but it does not ensure elimination of the infectious agent. Indeed, it has been shown that the residue within a burial site can persist for many years (NABC 2004) such that the ultimate elimination of carcases remains a long-term process.

The use of trench burial for carcase disposal was adopted during the 1984 AI outbreak in Virginia, US (Mixston 2003), during which 5,700 tons (5,170,953 kg) of carcase material were disposed of, with an estimated cost of $US25 per ~1000 kg (Berglez 2003). On-site burial was the primary method used and accounted for approximately 85% of the disposed carcases. Towards the end of the outbreak, the burial trenches were standardised at a width of 20 ft (6 m), a depth of 10 ft (3 m) and a length able to accommodate the carcases. This meant approximately 20 ft^3 were required per 800 lbs (about 363 kg) of poultry carcases.

12.6.1.2 Landfills

Landfills have been widely used as a means of carcase disposal in many disease eradication efforts, such as the 1984 and 2002 AI outbreaks in Virginia (Berglez 2003) and the 2002 outbreak of ND in southern California (Riverside County Waste Management Department 2003). The advantages of this method include:

- Landfill sites may be licensed to accept animal waste, hence dual purpose
- On-site facilities
- Large capacity
- Already existing and immediately available
- Environmental protection measures have been already designated and implemented

Among the disadvantages of landfills are:

- They may not be close to the source of the waste to be disposed of, thus risking the spread of disease agents during the transport of infected carcases (common to any off-site disposal methods).
- Commitment to site maintenance is long-term and hence expensive over an extended period.
- The process does not produce a usable by-product.
- The primary by-products resulting from decomposition of wastes in the landfill are leachate and landfill gas.

Leachate is defined as "liquid that has passed through or emerged from solid waste and contains soluble, suspended, or miscible materials removed from such waste" (US EPA 1995). The amount of leachate generated depends on the amount of liquid originally contained in the waste (primary leachate) and the quantity of precipitation that enters the landfill through the cover or that falls directly on the waste (secondary leachate) (US EPA 1995). The composition of leachate depends on the decomposition phase (acetic vs methanogenic phase). If the leachate is not properly managed, it can be released from the landfill and will result in environmental pollution. Landfill gases, typically 50% methane and 50% carbon dioxide, are the products of the anaerobic decomposition of organic material in landfill sites. If left unmanaged, landfill gas can vent to the atmosphere or migrate underground. Active control systems that rely on gas recovery wells or trenches and vacuum pumps to check the migration of landfill gas have been employed.

Modern Subtitle D landfills are designed to prevent the leakage of leachate from the site. The key features of these landfills include a composite liner, leachate containment systems and gas collection systems.

During the 1984 AI outbreak in Virginia, approximately 15% of the poultry carcase material was disposed of in landfills (Berglez 2003). The landfill used at that time was an unregulated dump, making potential groundwater and surface water contamination an issue. The environmental concerns resulted in only limited use of the site. In the 2002 AI outbreak in Virginia, commercial landfills played a more important role. During that outbreak, 16,900 tons of carcases were disposed of, 85% in landfills (Berglez 2003). Transportation of the waste proved to be the main bottleneck.

In October 2002, an outbreak of ND was confirmed in a backyard flock in southern California and spread to other, mainly backyard, flocks. During eradication approximately 3,160,00 birds were depopulated from 2,148 premises. Landfills were the primary method used to dispose of the carcases. The cost was estimated at about $US40 per ton (Hickman 2003). During the outbreak, the Riverside County Waste Management Division developed a training video for landfill operators on how to properly handle potentially infected waste (Riverside County Waste Management Division 2003).

12.6.1.3 Mass Burial

A large number of carcases can be accommodated in mass burial sites, which incorporate systems to collect, treat and dispose of leachate and gas. Mass burial sites played a key role in the 2001 outbreak of foot-and-mouth disease (FMD) in the UK and much of the information on this method was gained from that event. As shown by the UK experience, to minimise operational difficulties it is crucial that a site assessment is carried out prior to the initiation of site development. The total amount of land required depends on the volume of carcases and the space needed for operational activities. The most important advantage of mass burial is the capacity to dispose of a large number of carcases. However, the UK experience generated negative reactions to this method by the public. Among the disadvantages of mass burial, long-term costly monitoring and management of the facilities are the major issues.

12.6.1.4 Additional Remarks

For all the burial techniques described, the location of the sites should be recorded accurately. Site selection must include the following considerations: access to the site; environment (water table, proximity of municipalities, etc.) and construction (stability of soil, necessity of fencing and banks, etc.) (Ausvetplan 2007). Moreover, regular inspection of the burial site is recommended, with the aim of preventing problems and to return the site to its original condition. Correct site selection will affect the amount of time required for buried animal carcases to decompose as this depends on temperature, moisture and burial depth as well as on soil type and drainage.

The environmental impact of livestock burial has been poorly investigated (Freedman & Fleming 2003) and further studies are needed. The main environmental impact of mass burial is associated with the risk of potential contamination of groundwater with the chemical products of carcase decay. With reference to burial techniques of birds, two reports have provided evidence for these occurrences. The amount and type of contaminants released from two shallow pits containing 62,000 lbs of turkey carcases were evaluated by Glanville (1993, 2000). High levels of ammonia, total dissolved solids, biochemical oxygen demand (BOD) and chloride were observed in the monitoring well closest to the burial site. Studies by Ritter and Chrinside (1995, 1990) considered the impact of dead-bird disposal pits on groundwater quality. Over a 3-year monitoring period, some pits had impacted groundwater quality, with nitrogen being a greater problem than bacterial contamination.

12.6.2 Incineration

Historically, incineration played an important role in the disposal of carcases. However, increased awareness of public health issues and advances in technology have resulted in a reduction of its use. There are three categories of incineration techniques: (1) open-air burning, (2) fixed-facility incineration and (3) air-curtain incineration (NABC 2004).

12.6.2.1 Open-Air Burning

The burning of carcases in the open air, including on combustible heaps known as pyres, has been replaced by other disposal methods in many countries. The volume of ash produced can be massive (NAO 2002), with the potential for groundwater and soil contamination by the hydrocarbons used as fuel (Crane 1997).

Open-air burning is not permitted in every country or region and in most cases permission by local authorities has to be obtained. In a declared animal carcase disposal emergency, it may be possible to overcome local policy (Ellis 2001). Open-air burning is time-consuming and can be considered the most lengthy of the three incineration processes. The species of animal burned influences the length of the process. According to Berglez (2003), the greater the percentage of animal fat, the more efficiently a carcase will burn.

Open-air burning causes significant public awareness, often generating a negative image of the management of an outbreak. It is crucial during the site selection process to first communicate with local communities about open-air burning intentions (Widdrington FMD Liaison Committee).

12.6.2.2 Fixed-Facility Incinerators

These include small on-farm incinerators, small and large incineration facilities, crematoria and powder plant incinerators (NABC 2004). In contrast to open-air burning, the use of fixed-facility incinerators allows highly controlled and contained disposal. Fixed-facility incinerators are fuelled generally by diesel, natural gases or propane. Many incinerators are fitted with afterburner chambers that burn hydrocarbon gas completely. Compared to open-air burning, the ash produced is considered safe and may be disposed of in landfills (Ahlvers 2003). Fixed incinerators are more suitable for the disposal of small amounts of material and their lack of mobility results in their practicability being compromised (Ausvetplan 2007).

12.6.2.3 Air-Curtain Incineration

A relatively new technology for carcase disposal is air-curtain incineration. Here, a fan forces a mass of air through a manifold, creating a turbulent environment in which incineration is greatly accelerated, up to six times faster than open burning (NABC 2004; Ford 1994). The fans deliver high-velocity air down into either a metal refractory box or burn pit. Materials needed for the air-curtain system include wood (e.g. pallets in a wood-to-carcase ratio varying between: 1:2 and 2:1) fuel for the fire and an air-curtain fan (Ford 2003). Air-curtain facilities can vary in size and be constructed as mobile units. Other advantages are that they are designed to achieve high temperatures, resulting in an extremely efficient combustion, yielding better fire control and fuel economy than obtained with pyres (Ausvetplan 2007). However, they require active monitoring during operation and there must be a suitable location available in which to construct the pit.

12.6.2.4 Additional Remarks

Experience has shown that some disadvantages may be encountered during incineration, such as operation during atmospheric inversions (daily and weather front related); this has resulted in hanging smoke and odour and the high potential for equipment fires and other malfunctions. Immediate sources of back-up equipment should be identified and extensive air monitoring is necessary to ensure the safety of local residents (Flory et al. 2006). All of the methods described pose a fire hazard and yield ash.

12.6.3 Composting

Composting is a natural process during which microorganisms decompose biological material in the presence of oxygen, transforming the material into a safe and stable product (Ausvetplan 2007; NABC 2004; Mukhtar et al. 2004). Aerobic composting has been shown to be a valuable disposal technology. Carcase composting offers several advantages—from a reduced environmental impact to the generation of a valuable by-product and the destruction of pathogens.

The process of composting consists of two phases. During the first phase, the temperature increases, soft tissues decompose and bones begin decomposition. This phase may last from 3 weeks to 3 months (Haug 1993). In the second phase, decomposition of the remaining material, mainly bones, occurs. The compost turns into a black soil (humus) containing primarily nonpathogenic bacteria and plant nutrients. This phase takes approximately one month. The end of the second phase is marked by an internal temperature of 25–30°C. For this phase it is necessary to move the composting pile from a primary to a secondary bin.

In the composting of animal carcases, microorganisms convert the body of the dead animal and carbon source into a stable mixture of bacterial biomass and organic acids (Keener et al. 2000). Carcase composting systems need and rely on the availability of carbonaceous material. Carbon sources can include poultry litter, manure, cereal crop straw and other by-products such as peanut pods. Several ratios of carbonaceous material and animal waste are recommended in the literature. The Ausvetplan (2007) plan recommends a ratio of about 3:1 (w/w). According to NABC (2004), a 50:50 (w/w) mix can be used as a base for composting. A general rule is to define the ratio according to that of the carbon to nitrogen ratio (C:N). A ratio of carbon source materials to animal waste of 1:1 has been proposed for high C:N materials such as sawdust, 2:1 for medium C:N materials such as litter, and 4:1 for low C:N materials such as straw (NABC 2004). Table 12.3 summarises the recommended conditions for an active composting.

Bulking agents are also used during the composting process as they provide nutrients for the system and maintain adequate air space (25–35% porosity) within the compost pile by preventing the packing of the materials. The proposed ratio of bulking agent to carcases should result in a bulk density not exceeding 600 kg/m^3.

While the criteria guiding site selection vary depending upon local legal requirements, some characteristics should always be taken into account during the selection process. A compost site should be located in a well-drained area at least 90 cm above the water-table level and at least 90 m from water resources. It should also have an adequate slope (1–3%) that allows proper drainage. Runoff from the composting facility should be collected and directed away from production facilities (NABC 2004).

12.6.3.1 Windrow and Bin Composting

These two composting techniques share common guidelines even though different management principles may be required.

Windrow composting (Fig. 12.1) should be placed at the highest point on the identified site. A plastic liner covering the base of the windrow is needed as a moisture barrier. The liner should then be covered completely with co-composting material (such as sawdust or straw) to a thickness of about 30 cm for small carcases. A layer of bulking material (litter) is then placed on top to absorb moisture from the carcases and to maintain adequate porosity. The thickness of the bulking material should be 0.5 ft (15 cm) for small carcases (NABC 2004). A layer of carcases should be placed on top of the bulking material layer. In the case of small carcases, the first layer of animals can be covered with co-composting material and then a second layer of carcases placed over it. After the layering process, the entire windrow should be covered with a thick layer of biofilter material (carbon sources/bulking agents). With this construction method the approximate dimensions of the completed windrow for small carcases are: bottom width 3.6 m, top width 1.5 m, height 1.8 m.

Bin composting is well-suited to the disposal of small carcases. The required bin capacity will depend on the type of co-composting material used. Approximately 10 m^3 of bin capacity is required for 1,000 kg of carcases. Bins can be built with any material, such as wood or concrete. A simple and economical way to construct a bin is to use large round bales placed end to end to form a three-sided structure (also called bale composters). Bins may or may not be covered with a roof although it may be efficacious in rainfall areas, thereby reducing the potential for leaching from the pile. If bin walls are made of concrete, the recommended thickness is 15 cm. The height should be 1.5–1.8 m and the width should not exceed 2.4 m. The front of the bin is designed to

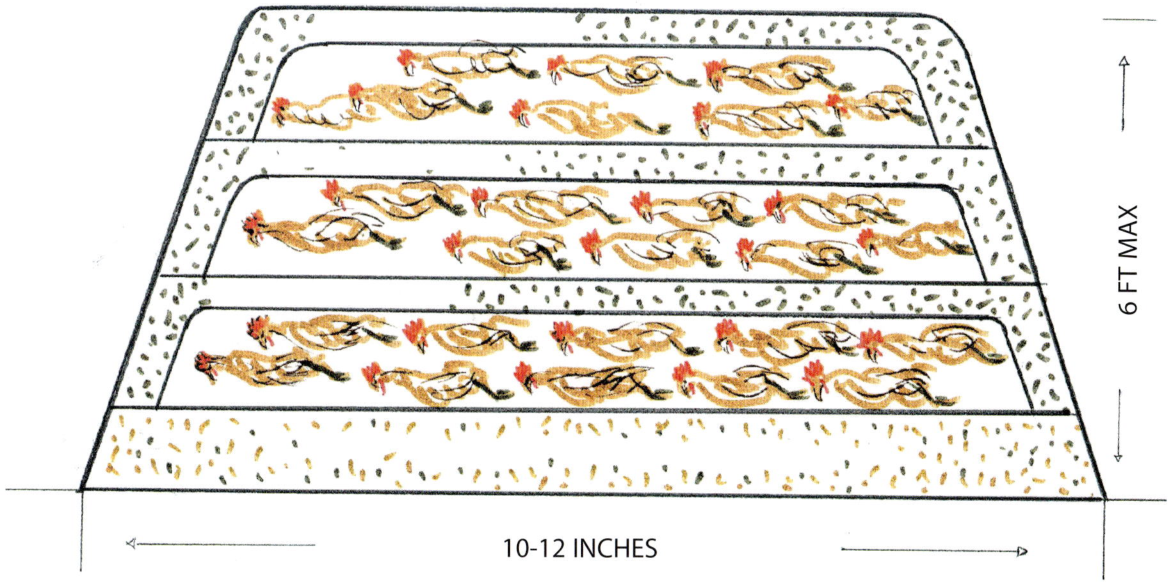

Fig. 12.1 Cross-section of carcass composting in a windrow (Carr et al. 1998). If straw is used, place 3-4 inches on top of sawdust or litter. Amount of sawdust can be reduced to 4-6 inches. (Courtesy of Amelio Meini)

Table 12.3 Poultry mortality rates and design weights (adapted from OSUE, 2000)[a]

Poultry Species and stage average	Weight in kg (lb)[a]	Poultry loss rate (%)[c]	Flock life (days)	Design weight IN kg (lb)[d]
Broiler	1.8-3.6 (4-8)	4.5-5	42-49	Up to 3.6 (up to 8)
Layers	2.0 (4.5)	14	440	2.0 (4.5)
Breeding hens	1.8-3.6 (4-8)	10-12	440	3.6 (8)
Turkey, females	6.8-11.4 (15-25)	6-8	95-120	11.4 (25)
Turkey, males	11.4-19.1 (25-42)	12	112-140	15.9 (35)
Turkey, breeders replace	6.8; 0-13.6 (15; 0-30)	5-6	210	9.1 (20)
Turkey, breeding hen	12.7-13.6 (28-30)	5-6	180	13.6 (30)
Turkey, breeding tom	31.8-36.4 (70-80)	30	180	34.1 (75)

[a]From NABC (2004).
[b]Average weight used to calculate pounds of annual mortality.
[c]For mature animals, the percent loss is an annual rate for the average number of head on the farm.
[d]Design weight used to calculate composting cycle periods.

allow easy loading of carcases, thus ensuring that the carcases are not raised over a height of more than approximately 1.5 m.

The bin composting process comprises two phases: the first one is essentially a litter base (40–50 cm) that is placed in the bin 2 days before the carcases to allow pre-heating of the litter. Prior to introduction of the carcases, 15 cm of the pre-heated litter should be removed and the carcases placed on the remaining litter. This will absorb any fluids, preventing leakage. The carcases are then covered completely with of the remaining pre-heated litter. Next, the carcases are layered, placing a thick cover of carbon source material between carcase layers. A final cover (60 cm) of sawdust should be added on top. The second phase of the process involves moving the pile to a secondary bin, which is then covered with a minimum of co-composting material. Moisture is added to the material to allow the pile to reheat, as this is essential for an acceptable end product.

For successful composting, time, porosity, aeration and, especially, temperature are crucial factors. Although high compost temperatures promote rapid decomposition and effective pathogen elimination, excessively high temperatures may inactivate desirable enzymes. The time needed to complete the composting process depends on a variety of factors. Generally, composting time is shorter in warmer climates than in colder ones. The size of the animal also affects the time required. The estimated time at which piles are moved from the primary to the secondary phase for small carcases such as poultry is 7–10 days (NABC 2004). Murphy and Carr (1991) reported that the composting of broiler carcases required two consecutive 7-day periods to reduce carcases to bony residues. Appendix C contains a description of the calculation of the correct design parameters for an effective composting facility for poultry. Murphy and Carr (1991) and Keener and Elwell (2000) developed a model based on a mathematical formula for the calculation of composting volumes. Tables 12.3 and 12.4 describe the design weight used to calculate composting cycle periods for poultry.

Table 12.4 Recommended conditions for active composting (Rynk 1992)

Carbon-to-nitrogen (C:N)ratio[b]	20:1–40:1
Moisture content	65%
Oxygen concentration[c]	>5%
Particle size (diameter in inches)	0.5–2
Pile porosity	>40%[d]
Bulk density	474–711 kg/m^3 (800–1,200 lb/yd^3)
pH	5.5–9
Temperature (°F)	110–150

[a]Although these recommendations are for active composting, conditions outside these ranges may also yield successful results.
[b]Weight basis (w:w). C:N ratios > 30 will minimise potential odours.
[c]An increasing likelihood of significant odours occurs at approximately 3% oxygen or less. Maintaining oxic conditions is key to minimising odours.
[d]Depending upon the specific materials, pile size and/or weather conditions.

12.6.3.2 Additional Remarks

Lessons learned from the 2002 AI outbreak in the US suggest that factors important for successful in-house composting are the active involvement of poultry companies in managing the process, the formation of an expert team, the availability of carbon material and litter, identification of sources of carbon and water and the rapid identification of response teams that are trained and equipped to compost flocks within 24 h of virus confirmation (Flory et al. 2006). According to the US experience, in-house composting is preferred to the other disposal methods. The main disadvantages of this method include the need for long-term management and the risk incurred by the need to transport the carbon material off-site, with a potential risk of secondary spread of the infectious agent by transport vehicles. Under certain atmospheric conditions, an unpleasant odour may persist for extended periods of time.

12.6.4 Rendering

The process of cooking and sterilising non-edible waste is referred to as rendering. Specifically, it has been defined as the separation of fat from animal tissues by the application of heat (NABC 2004) with the goals of eliminating water, sterilising the final products, and producing meat and bone meal (MBM) from dead animals or waste materials associated with slaughtering operations (Kamur 1989). The meat meal derived from rendering poultry waste is technically referred to as poultry by-product hydrolysed feather meal (PBHFM), or simply meat meal. Meat meal is 60% protein and 20–22% fat.

Rendering involves the use of high temperatures and pressure to convert animal carcases to safe, nutritional and valuable economic products (UK-DEFRA 2000). Animal carcases are converted into three main end products by rendering: carcase meal, melted fat and water. The main procedures for rendering carcases involve size reduction, cooking and the separation of fat, water and protein materials. Rendering processes may be divided into "edible" or "inedible" (NACB 2004). During edible rendering, the carcase by-products are reduced into small pieces and disintegrated by cooking, resulting in moisture and edible tallow or fat. Inedible rendering converts protein, fat and keratin materials into tallow, carcase meal (used in livestock feed, soap, production of fatty acids) and fertilizer, respectively. Raw materials are dehydrated and cooked, and fat and protein subsequently separated. The two rendering processes differ in their raw materials and end products. Several rendering systems exist, two of which are summarised below.

"Wet rendering" adds moisture to the raw materials during the cooking process. Although this method produces good-quality tallow, it is less frequently used because of its high energy consumption and adverse effect on fat quality (Ockerman and Hansen 2000). It has been reported that the accumulated water in this system needs extra energy to evaporate, with the consequence of material remaining, termed "sticky liquor" (Romans et al. 2001).

The newer method of "dry rendering" uses heat generated by steam condensation and applied to agitator blades to obtain uniform heat distribution. This shortens the time needed to cook the carcases. The indirect heat applied in this system converts the moisture in the carcases to steam. During this process, the yield of meat meal is higher than that obtained by wet rendering.

Both systems, wet and dry, can be converted into a batch system consisting of multiple cooker units (usually two to five). Most rendering options have a continuous cooker such that all rendering steps are carried out simultaneously and consecutively (EPAA 2002). The system needs little or no manual operation and end products are generated at a constant rate using indirect steam.

The time required for a rendering process depends mostly on the temperature and air pressure. Increasing both factors decreases automatically the rendering time. Air pressure mainly impacts the quality of the outgoing products. The advantages and disadvantages of the rendering process were summarised by Flory et al. (2006).

Advantages:

- The poultry industry owns some of the rendering plants, giving it more control over the disposal process.
- Long-term management is not required.
- No environmental impact.
- Produces a usable end product (market uncertain).
- If no market for the product exists, rendered proteins can be transported biosecurely to the landfill.

Disadvantages:

- Rendering plants are often located close to poultry operations; thus, all possible sources of disease transmission must be identified and controlled.
- Plant capacity may not be adequate.

- Due to upgraded biosecurity requirements, a plant may need to be dedicated to rendering AI carcases for the duration of the outbreak. This may not be economically feasible for a limited outbreak.
- Integrators without rendering capability would be at the mercy of a private rendering company.
- Rendering costs are uncertain and can dramatically increase during an outbreak.

The following must be considered in the determination of whether or not rendering is the most suitable method for the disposal of birds infected with the AI virus:

- Discussions with rendering companies and the poultry industry should take place before the occurrence of the outbreak.
- Most rendering facilities are privately owned (i.e. not owned by the poultry industry) and are not allowed to accept material infected with AI or ND virus.

Appendix A: Practicalities of Decontamination with Formaldehyde Gas

Formaldehyde gas can be used with safety only in certain environments and in the hands of experienced operators. Effective decontamination with gaseous formaldehyde requires a favourable combination of gas concentration, temperature, relative humidity and contact time. Most procedures suggest formaldehyde concentrations of 2–10 g/m^3 and a relative humidity of 70–90% at temperatures of 20°C for periods of 15–24 h. Electric fans, where present, should promote homogeneous dispersal of the gas in the enclosed space. Although a high relative humidity is necessary for optimal activity, water cannot be present in liquid form, as it will dissolve the gas and reduce its concentration in the gaseous phase. It is therefore difficult to establish the required relative humidity conditions outside a controlled laboratory situation. An evenly controlled temperature is also essential for effective decontamination. If the temperature of the walls of the vessel or building falls during decontamination, the formaldehyde will polymerise on them to form a powdery precipitate of paraformaldehyde, which reduces the effectiveness of the operation and creates problems of residual toxicity. Such conditions are likely to occur in farm buildings or vehicles during overnight decontaminations.

Fumigation should be adopted at the end of the disinfection procedure to optimise the viricidal effect of the formaldehyde gas. To produce the gas, formalin solution (20 ml/m^3 space) can be added to potassium permanganate (16 g/m^3); a violent reaction that produces heat and boiling will follow and is potentially dangerous to the inexperienced operator. The enclosure must be prepared in advance so the operator, wearing protective clothing and a full facial respirator, can mix the ingredients and leave the enclosure quickly. Because formaldehyde is a very toxic gas, it must be totally retained within the space to be treated and then effectively neutralised prior to exposure, by reaction with ammonia gas obtained from the heating of ammonium carbonate. Breathing masks and special equipment for monitoring residual formaldehyde are essential.

Appendix B: Techniques of Humane Destruction of Animals

When selecting a killing method, only those that can guarantee a high-volume killing capacity under all weather circumstances should be used. All birds to be killed for disease control purposes should be handled with the same care and concern for their welfare as those that are killed for food. Killing for disease control purposes and vaccination should be carried out only by properly trained individuals. Training should be provided at times when there is no disease outbreak so that efficient, trained persons are available when an outbreak occurs. Resources should be made available to create a group of trained facilitators for emergency culling of large numbers of birds. It is advisable to involve the local farming community in drawing up plans for each farm or type of farm during non-crisis times, so that in the event of an outbreak of a disease such as AI there will be an optimal killing process with a minimal amount of animal suffering. Birds vary considerably in their size structure and physiology. Since many species require expert handling during euthanasia, it is recommended that careful planning and consultation be carried out first. If there is a risk that the virus will spread to wild or captive birds, the welfare of these birds should be preserved.

Generally, carbon dioxide (CO_2) gassing or barbiturate overdose are the methods of choice for euthanasia. For small numbers of birds (e.g. fancy breeds and pigeons), the preferred method may be dislocation of the neck (using forceps or bare hands) or the injection of barbiturate.

Euthanasia of cassowaries, emus, ostriches, brolgas and other unusual/difficult birds requires expert assistance. The preferred options for large birds are lethal injection (for managed birds) and firearms (for free-ranging birds). Developing embryos in fertilised eggs can be killed by cooling them to +4°C for 4 h. For large numbers of birds in commercial poultry units, the preferred method is gassing with CO_2. Birds can be caught by teams of 10–15 labourers (experienced catching teams are preferable). Chicks are easily caught under heaters and are transferred in plastic garbage bins to waste skips for CO_2 gassing. Broilers on the ground can be driven to the catching area, where they can be caught and then placed directly into skips.

Caged birds are more difficult and progress will be slower. Skips should be filled to a level (between 70 and 90%) such that the remaining CO_2 gas layer will effectively kill the last layer of birds and the truck is not overloaded. The skip is then sealed and transported to the disposal site. Care must be taken to ensure that no bird is still alive when dropped into the burial pit.

The following methods of killing poultry for AI control are recommended by the Animal Health and Welfare Panel of the European Food Safety Authority (EFSA 2005).

- The birds are placed in suitable containers, including effectively restricted areas of a building, containing appropriate inert gas mixtures, such as argon, with not more than 2% oxygen.
- The birds are put into a suitable container of pure 4–6% carbon monoxide gas for a duration of at least 6 min; proper safeguards for human operators must be implemented.
- With the exception of ducks and geese, for which CO_2 should not be used, birds are exposed to not more than 30% CO_2 in an inert gas, such as nitrogen or argon, and not more than 2% oxygen.
- The use of a portable electrical stunner, poultry killer or captive bolt stunner is allowed but only if death can be confirmed in each animal.
- Individual birds can be injected with barbiturates; this method is impractical for large numbers of birds. For poultry during the first week of life, the chicks may be dropped into a macerator, which kills the bird instantaneously.

Other methods, such as putting birds into plastic bags and burning them or gassing them with hydrogen cyanide, impure carbon monoxide, or high concentrations of CO_2, are not allowed, neither is the gassing of whole buildings without adequate restriction of the area occupied by the gas or injection with any chemical except barbiturates.

Appendix C: Design Parameters for an Effective Poultry Composting Facility

The formula presented by Murphy and Carr (1991) is based on the concept that the capacity of bin systems for composting poultry depends on the theoretical farm live weight. The authors described a model in which the peak capacity of dead poultry for the first phase of composting is predicted based on the market age and weight of birds (see example 1 below):

Daily composting capacity = theoretical farm live weight/400

Theoretical farm live weight = farm capacity × market weight

Keener and Elwell (2000) developed models based on the results of experiments for a bin system for poultry (broilers). They assigned a specific volume coefficient of 0.0125 m^3/kg mortality/growth cycle (0.20 ft^3/lb mortality/growth cycle) for calculating primary, secondary and storage volumes (V1, V2, and V3, respectively). As discussed earlier, the composting times of primary, secondary and storage phases (T1, T2 and T3, respectively) are affected by various factors in the composting pile and are not equal to each other. Based on the above information, the authors suggested the following models for calculating the composting time and volume needed for primary, secondary and storage phases:

T1 = (7.42) (W1) 0.5 ≥ 10, days (5) V1 ≥ (0.0125) (ADL) (T1), m^3

T2 = (1/3) (T1) ≥ 10, days (7) V2 ≥ (0.0125) (ADL) (T2), m^3

T3 ≥ 30, days (9) V3 ≥ V2 or V3 ≥ (0.0125) (ADL) (T3), m^3

where W1 is the average weight of mortality in kg, and ADL is the average daily loss or rate of mortality in kg/day.

Example 1: Bin Composting of Poultry Carcasses Calculation

The following example is based on the method of Murphy and Carr (1991).

Available Information

- A poultry farm with 100, 000 birds of 4.5 lb (2.02 kg) average market weight where carcases are to be composted using a bin system.
- 0.45 kg (1 lb) of the compost material needs a volume of approximately 0.027 m^3 (1 ft^3).
- Daily composting capacity = theoretical farm live weight/400.
- Theoretical farm live weight = farm capacity × market weight.

Daily Composting Capacity

Daily composting capacity =100,000 (birds) × 4.5 (lb/birds)/400 (day) = 1125 lb/day (506.25 kg/day) or about 1125 ft^3/day

Suggested Number of Bins and Their Dimensions

Based on the experimental data of Murphy and Carr (1991), the most appropriate bin dimensions are 7 ft length, 5 ft width and 5 ft height. Therefore:

- N (number of primary treatment bins) = (compost capacity)/(L × W × H of a primary bin).
- N = (1,125 ft^3/day)/(7 ft × 5 ft × 5 ft) = 6 primary treatment bins/day.
- The six bins can be arranged in any of several configurations to suit the needs of a particular situation.
- Overall length = (1,125 ft^3)/(7 ft × 5 ft) = 32 ft (9.64 m).
- Total area = 7 ft × 32 ft = 214 ft^2 (19.26 m^2).
- Area for each primary bin= 214 ft^2/6 = 35 ft^2 (3.21 m^2).

Example 2: Bin Composting of Poultry Carcase Sample Calculation

The following example is based on the method of Keener and Elwell (2000).

Available Information

A poultry farm with an average weight of 1.36 kg (3 lb) per carcase and ADL of 13.6 kg/day (30 lb/day where carcases are to be composted using a bin system.

T1 = (7.42) (W1) 0.5 ≥ 10, days V1 ≥ (0.0125) (≥ ADL) (T1), m^3

T2 = (1/3) (T1) ≥ 10, days V2 ≥ 0.0125) (ADL) (T2), m^3

T3 ≥ 30, days V3 ≥ V2 V3 ≥ (0.0125) (ADL) (T3), m^3

The relation between bin volumes, width, and length with a constant depth or height of 1.50 m (5 ft).

Composting Time and Volume for Primary, Secondary and Storage Phases

From the above equations, the required information is:

T1 = (7.42) (1.36) 0.5 ≥ 10 days, T2 (1/3) (T1) ≥ 10 days and T3 ≥ 30 days,

V1 ≥ (0.0125) (≥13.6) (10) =1.70 m^3, V2 ≥ 0.0125) (13.6) (10) = 1.70 m^3 and

V3 ≥ 3 V2 (recommended as a design parameter) = 3 (1.70) = 5.10 m^3.

Number of Required Bins and Their Associated Dimensions

The bin volume closest to a calculated value of 1.70 m^3 is 2.26 m^3 (80 ft^3) or a mini-bin with dimensions of 1.22 m × 1.22 m × 1.52 m (4 ft × 4 ft × 5 ft).

Thus, two primary bins, each with an area of 1.22 m × 1.22 m =1.5 m^2 (16ft^2) or a total of 3 m^2 (32 ft^2), and one secondary bin of 1.50 m^2 (16 ft^2) are needed.

The end-product storage area is 5.10 m^3/ 1.5 m = 3.36 m^2.

References

Ahlvers D (2003) Personal communication to Justin Kastner regarding Kansas State University College of Veterinary Medicine incinerator: Dennis Ahlvers (Physical Plant Supervisor, Kansas State University)

Ausvetplan Australian Veterinary Emergency Plan, Version 3.1, 2006. Disease Strategy. Avian influenza. Interim Draft

Ausvetplan Australian Veterinary Emergency Plan, Version 3.1, 2006. Disease Strategy. Newcastle disease

Ausvetplan Australian Veterinary Emergency Plan. Operational Procedures Manual Disposal, Version 3.0, 2007. Carcass Disposal: A Comprehensive Review National Agricultural Biosecurity Center Consortium USDA APHIS Cooperative Agreement Project Carcass Disposal Working Group, August 2004

Ausvetplan Australian Veterinary Emergency Plan. Operational Procedures Manual, Version 3.0, 2006. Destruction of animals: A manual of techniques of humane destruction

Ausvetplan Australian Veterinary Emergency Plan. Operational Procedures Manual, 2000. Decontamination

Ausvetplan Australian Veterinary Emergency Plan. Persistence of Disease Agents in Carcases and Animal Products. Report for Animal Health Australia, by Scott Williams. Consulting Pty Ltd Revised - December 2003

Beard CW, Hanson HP (1984) Newcastle Disease. In: Diseases of poultry. Hofstad MS, Bames HJ, Calnek BW, Reid WM, Yoder HW (eds), 8th ed, Ames, IA: Iowa State University Press, pp. 452-470

Berglez B (2003) Disposal of poultry carcasses in catastrophic avian influenza outbreaks: A comparison of methods (technical report for Master of Public Health). Chapel Hill: University of North Carolina

Carr L, Broide HL, John HM et al (1998) Composting catastrophic event poultry mortalities. Maryland: University of Maryland & Maryland Cooperative Extension. Retrieved April 21, 2003, from http://www.agnr.umd.edu/MCE/Publications/Publication.cfm?ID=35

Crane N (1997) Animal disposal and the environment. State Veterinary Journal 7(3):3-5

Davison S, Benson CE, Ziegler AF, Eckroade RJ (1999) Evaluation of disinfectants with the addition of antifreezing compounds against non pathogenic H7N2 avian influenza virus. Avian Dis 43(3):533-537

De Benedictis P, Beato MS, Capua I (2007) Inactivation of avian influenza viruses by chemical agents and physical conditions: a review. Zoonoses Public Health 54(2):51-68

EFSA (2005) Animal health and welfare aspects of Avian Influenza. The EFSA Journal 266:1-21

Ellis D (2001) Carcass disposal issues in recent disasters, accepted methods, and suggested plan to mitigate future events (applied research project for Master of Public Administration). San Marcos, Texas: Texas State University - San Marcos (formerly Southwest Texas State University)

EPAA European Partnership for Alternative Approaches to Animal Testing (2002) Abattoirs: air emission control. Retrieved June 19, 2003, from http://www.epa.nsw.gov.au/mao/abattoirs.htm

Flory GA, Peer RW, Bendfeldt ES (2006) Evaluation of Poultry Carcass Disposal Methods Used During an Avian Influenza Outbreak in Virginia in 2002. Virginia Department of Environmental Quality

Ford G (2003) Disposal Technology Seminar on Air-Curtain Incineration. Kansas City, Missouri: Midwest Regional Carcass Disposal Conference

Ford WB (1994) Swine carcass disposal evaluation using air curtain incinerator system, model TCh.2. ?Incineration 25:359 (Foreign Animal Disease Report, 22-2; reprinted by Air Burners, LLC, at http://www.airburners.com/DATAFILES_Tech/ab_swine_report.pdf). Washington: USDA Animal and Plant Health Inspection Service

Freedman R, Fleming R (2003) Water quality impacts of burying livestock mortalities. Livestock Mortality Recycling Project Steering Committee, August 2003, Ridgetown, Ontario, Canada. Available at http://www.ridgetownc.on.ca/research/documents/fleming_carcassburial.pdf

Glanville TD (1993) Groundwater impacts of onfarm livestock burial. Iowa Groundwater Quarterly 4:21-22

Glanville TD (2000) Impact of livestock burial on shallow groundwater quality. Paper presented at ASAE Mid-Central Meeting, St. Joseph, Missouri (No. MC00-116)

Haug RT (1993) The practical handbook of compost engineering. Boca Raton, Florida: Lewis Publishers, Press, Inc

Hickman M (2003) Disposal of END waste in southern California by landfill. Kansas City, Missouri: Midwest Regional Carcass Disposal Conference

Keener HM, Elwell DL, Monnin MJ (2000) Procedures and equations for sizing of structures and windrows for composting animal mortalities. Applied Engineering in Agriculture 16(6):681-692

Mukhtar S, Auvermann BW, Heflin K, Boriack CN (2003) A low maintenance approach to large carcass composting. Paper written for presentation at the 2003 ASAE Annual International Meeting, Las Vegas, Nevada. Paper Number: 032263

Murphy DW, Carr LE (1991) Composting dead birds. Extension Sections in Departments of Poultry Science and Agricultural Engineering, University of Maryland at College Park. Fact sheet 537. Retrieved on April 11, 2003 from http://www.agnr.umd.edu/MCE/Publications/Publication.cfm?ID=145&cat=3

NABC National Agricultural Biosecurity Center Consortium. USDA APHIS Cooperative Agreement Project. Carcass Disposal Working Group, August 2004. Carcass Disposal: A Comprehensive Review

NAO National Audit Office (2002) The 2001 outbreak of foot and mouth disease. London: UK. National Audit Office

Noll H, Youngner JS (1959) Virus-lipid interactions. II. The mechanism of adsorption of lipophilic viruses to water-insoluble polar lipids. Virology 8(3):319-343

Ockerman HW, Hansen CL (2000) Rendering: Animal byproduct processing and utilization. Washington, DC: CRC Press LLC

Osue (2000) Ohio's livestock and poultry mortality composting manual. Columbus, Ohio, USA: The Ohio State University Extension, available at http://www.oardc.ohio-state.edu/ ocamm/Keener-Maine%20Mortality%20Paper%205-24-05.pdf

Ritter WF, Chirnside AE (1990) Dead bird disposal and ground-water quality. Proceedings of the 6th International Symposium on Agricultural and Food Processing Wastes, Chicago, Illinois, pp 414-423

Ritter WF, Chirnside AE (1995) Impact of dead bird disposal pits on ground-water quality on the Delmarva Peninsula. Bioresource Technology 53:105-111

Riverside County Waste Management Department. Landfill fees. Retrieved September 10, 2003, from http://www.rivcowm.org/landfill_fees_02.htm

Romans JR, Costello WJ, Carlson CW et al (2001) Packing house by-products. In: The Meat We Eat. Danville, Illinois: Interstate Publishers, Inc

Rynk R (1992) On-farm composting handbook. Ithaca, New York: Northeast Regional Agricultural Engineering Service

Samberg Y, Meroz M (1995) Application of disinfectants in poultry hatcheries. Rev Sci Tech 14(2):365-380

UK Environment Agency (2001) The environmental impact of the foot and mouth disease outbreak: an interim assessment

US EPA. 40 CFR Part 258 - Criteria for municipal solid waste landfills

US EPA. Office of Solid Waste and Emergency Response. (1995). Decision-makers' guide to solid waste management (Vol. 2). Washington

UKDEFRA United Kingdom Department for Environment Food and Rural Affairs (2000). The BSE Inquiry: The Report. Vol. 13: Industry Processes and Controls. Ch. 6: Rendering. Annex B: Manufacturing process of rendering. Retrieved June 10, 2003, from http://www.bseinquiry.gov.uk/report/volume13/chapterj.htm

Widdrington FMD, Liaison Committee. Submission to Northumberland FMD Inquiry

Websites

Organizations

World Organisation for Animal Health (OIE)
www.oie.int

- Update of Highly Pathogenic Avian influenza in Animals
- www.oie.int/downld/AVIAN%20INFLUENZA/A_AI-Asia.htm
- Avian influenza
  www.oie.int/eng/info_ev/en_AI_avianinfluenza.htm

United Nations Food and Agriculture Organization (FAO)
www.fao.org

- Preparing for Highly Pathogenic Avian influenza
  www.fao.org/docs/eims/upload//200354/HPAI_manual_en.pdf
- EMPRES (Emergency Prevention System for Transboundary Animal and Plant Pests and Diseases
  www.fao.org/ag/againfo/programmes/en/empres/home.asp
- Avian influenza
  www.fao.org/avianflu/en/index.html

The World Health Organization (WHO)
www.who.int

- Avian influenza
  www.who.int/csr/disease/avian_influenza/en/index.html

Center for Disease Control and Prevention (CDC)
www.cdc.gov

- Pandemic Flu
  www.pandemicflu.gov

European Community Animal Health and Welfare
www.europa.eu

- Avian influenza
  ec.europa.eu/food/animal/diseases/controlmeasures/avian/index_en.htm
- 2004/402/EC: Commission Decision of 26 April 2004 approving contingency plans for the control of avian influenza and Newcastle disease eur-lex.europa.eu/LexUriServ/LexUriServ.do?uri=CELEX:32004D0402:EN:NOT

I. Capua, D.J. Alexander (eds.) *Avian Influenza and Newcastle Disease*,

European Centre for Disease Prevention and Control (ECDC) www.ecdc.eu.int	• Avian influenza www.ecdc.eu.int/Health_topics/Avian_Influenza/Avian_Influenza.html
Center for Infectious Disease Research and Policy www.cidrap.umn.edu	• Avian influenza http://www.cidrap.umn.edu/cidrap/content/influenza/avianflu/index.html
European Food Safety Authority (EFSA) www.efsa.europa.eu	• Avian influenza and Food http://www.efsa.europa.eu/EFSA/efsa_locale-1178620753812_AvianAndFoodFAQs.htm
FLU in China and FLU Information Center (FIC)	• http://www.flu.org.cn/en/default.html

Networks

The OIE and FAO network of expertise on Avian influenza	http://www.offlu.net
FAO – CIRAD - Wetland international	http://wildbirds-ai.cirad.fr
FluTrop-Avian influenza research in Tropical countries	http://avian-influenza.cirad.fr
Global Avian influenza network for Surveillance	http://www.gains.org/

Genetic Databases

Influenza Virus Database	http://influenza.genomics.org.cn/index1.jsp
Influenza Virus Resource	http://www.ncbi.nlm.nih.gov/genomes/FLU/FLU.html
Global Iniziative on Sharing Influenza Data (GISAID)	http://www.gisaid.org

Email Alert Services

Avian influenza Intelink Digest	http://www.intelink.gov
ProMed-mail	http://www.promedmail.org

Annex 1
Check List for Visit to Suspect Premise

Manuela Dalla Pozza

Kit for the Official Veterinarian (OV)

1. Epidemiological inquiry form
2. Equipment necessary for the clinical visit and sampling procedures:
   a. 2 disposable suits
   b. 5 pairs of disposable shoe-covers
   c. 2 pairs of rubber gloves and 5 pairs of latex gloves
   d. disposable caps and face masks
   e. protective goggles
   f. paper tissues
   g. 5 leak-proof containers
   h. 5 leak-proof and water-resistant plastic bags
   i. electric torch
   j. disinfectant solution
   k. 2 pens and a notepad
   l. 100 2.5-ml syringes, with needles
   m. 100 thin, small plastic bags
   n. 2 pairs of surgical scissors
   o. 2 pairs of forceps
   p. tape
   q. 2 felt-tip pens
   r. 1 thermic container
   s. 5 frozen icepacks

   At least two of these kits should be prepared and available at the OV headquarters at all times.

Kit for the Laboratory Veterinarian

a. 1 thermic container
b. 4 pairs of forceps
c. 2 pairs of surgical scissors
d. 1 knife
e. tape
f. labels
g. 100 2.5-ml syringes with needles
h. sterile swabs
i. 50 test tubes containing virus transport media
j. 10 leak-proof containers
k. 2 disposable suits
l. 5 pairs of disposable shoe-covers
m. 5 pairs of latex gloves
n. disposable caps and face masks
o. protective eye goggles
p. 10 black waste-bags
q. 50 rubber bands
r. disinfectant solution in a nebuliser
s. cardboard container

I. Capua, D.J. Alexander (eds.) *Avian Influenza and Newcastle Disease*,

Annex 2
Epidemiological Investigation Form for Avian Influenza and Newcastle Disease Outbreaks

Manuela Dalla Pozza

Date ..../......../........

Dr .... Phone number ....

Suspicion registration N°. .... /Confirmation registration N° ....

Name of holding ....

Address ....

CAP/Zip/Postal code ....

Province/State/County .... Phone ....

Farm code or identification number □□□ □□ □□□□

Owner ....

Company ....

Address of the owner .... Phone ....

Information provided by ....

....

Farm Veterinarian Dr. .... Present NO ❑ YES ❑

HATCHERY OF ORIGIN

Company Hatchery NO ❑ YES ❑

Company .... Address ....

CAP/Zip/Postal code .... Province/State/County ....

Code □□□ □□ □□□□

Phone .... Fax ....

I. Capua, D.J. Alexander (eds.) *Avian Influenza and Newcastle Disease*,

Information Concerning the Farm

Type of Establishment	❑ Commercial intensive (indoor housing/controlled environment) ❑ Commercial extensive (access to outdoor facilities/natural environment) ❑ Independent rearer/trader ❑ Rural/Backyard
Category/Production line	Table-egg layers ❑ Type of production: Grandparents/primary breeder ❑ Layer Parents ❑ Rearing Pullets ❑ Layers ❑ All-in all-out system YES ❑ NO ❑
	Meat birds ❑ Type of production: Grandparents/primary breeder ❑ Broiler breeders parents ❑ Rearing pullets ❑ Meat-type (*broiler*) ❑

Species and number of birds present at the time of implementation of restriction measures

Type of production	Date of housing (restocking, placing)	No. of restocked birds		No. of birds alive at the time of implementation of restriction measures		Average weight of birds		No. of sick birds	No. of dead birds
		Female	Male	Female	Male	Female	Male		
Broiler									
Broiler breeder (all ages)									
Laying hen									
Meat turkey									
Turkey breeder (all ages)									
Guinea fowl breeder (all ages)									
Meat Guinea fowl									
Meat duck									
Duck breeder (all ages)									
Pigeon									
Pheasant									
Goose									
Quail									
Other species (specify)									
Other species (specify)									

Housing system

Sheds NO ❑ YES ❑ N°……...............
Tunnels NO ❑ YES ❑ N°…................

Feed supplier:
Feed heat treated NO ❑ YES ❑
Swill feeding NO ❑ YES ❑
If yes source of swill: …………………………………………………………………………………

Feed storage:
Bin ❑ Bag ❑ Loose bulk ❑

Type of ventilation system:
Natural ....
Natural with fans ....
Artificial ....

Free-ranging system NO ❑ YES ❑ m^2…..........
Bird-proof nets NO ❑ YES ❑

Possibility of contact with wild birds:
NO ❑ YES ❑ Species ....
....
....

Debeaking operations: Date ……/…../……
Performed by: Family members ❑ Employed staff ❑ External staff ❑ Other ❑ ……...............….
Remarks ....
....
....

Other birds present on site (captive or free-range)
NO ❑ YES ❑ Species ....
....
....

Presence of ponds
or lakes: NO ❑ YES ❑
Other water reservoirs NO ❑ YES ❑ (specify) ....

Water supply:
Mains NO ❑ YES ❑
Shallow well NO ❑ YES ❑
Deep well NO ❑ YES ❑
Lake group scheme (name): ……………………………………………

Water storage:
Outside NO ❑ YES ❑
Inside NO ❑ YES ❑
Covered NO ❑ YES ❑

Presence of pigs NO ❑ YES ❑ N° ....
Other animals NO ❑ YES ❑ (specify) ....
Remarks ....
....

1. Topography of the establishment

A map of the infected premises must be drawn, indicating clearly the productive units, the animals housed inside them and the main routes of access to the premises.

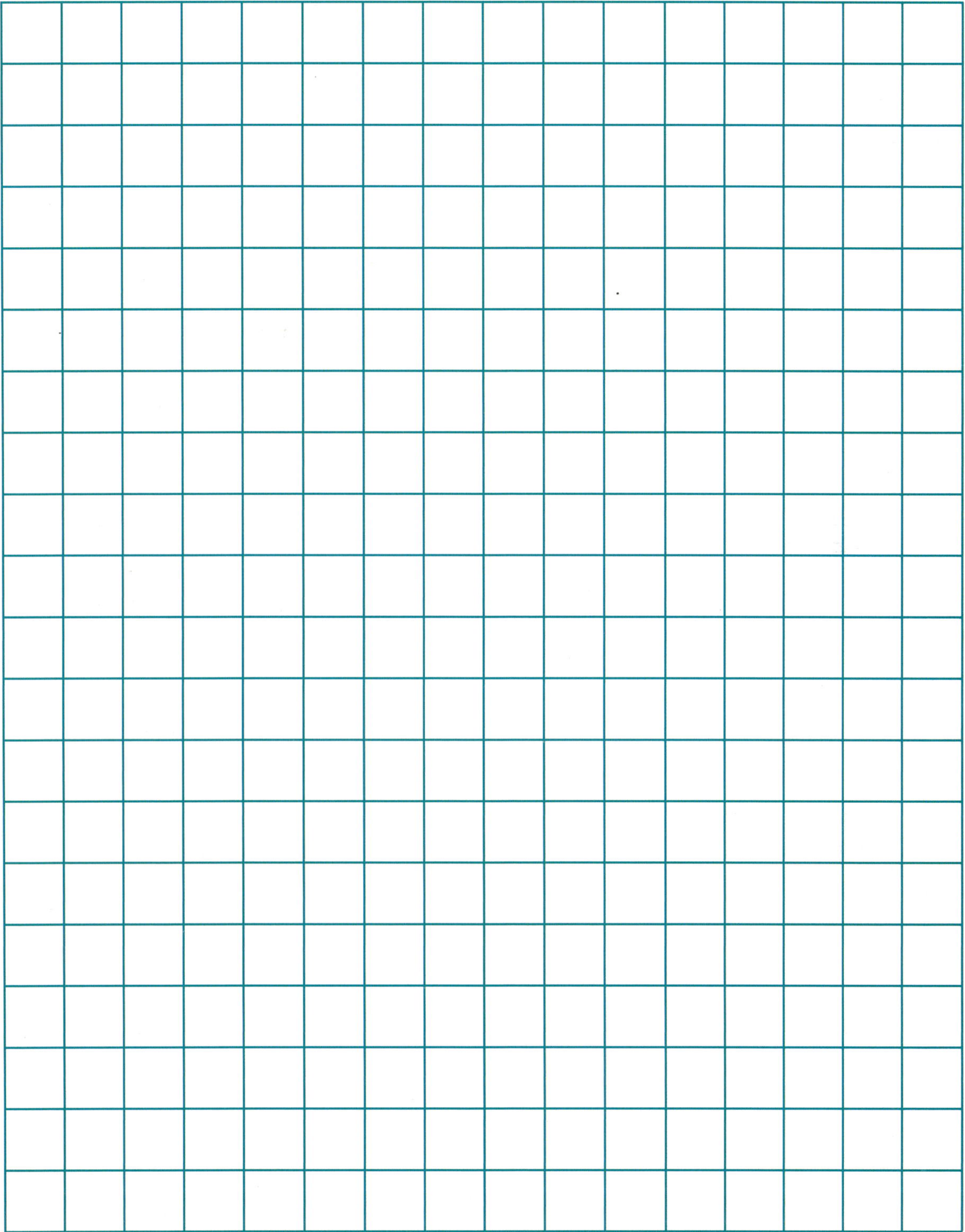

Data on introduction/spread of infection: information necessary for points 1–3, below must be collected for all movements of animals/people and should be repeated for each potential risk-associated event.

1. *Movements of birds*

a) Introduction of birds from other establishments/hatcheries/farms NO ❑ YES ❑

(From 21 days before the onset of the first clinical signs)

Date ..../..../.... N° .... Species .... Farm ❑ Hatchery ❑

Name of farm .... Code □□□ □□ □□□□

Address ....

CAP/Zip/Postal code .... Province/State/Country ....

Car plate of vehicle ....

---

b) Introduction of birds from exhibitions/markets/fairs NO ❑ YES ❑

(From 21 days before the onset of the first clinical signs)

Date ..../..../.... N° .... Species ....

Origin: Fair ❑ Market ❑ Exhibition ❑

CAP/Zip/Postal code .... Province/State/Country ....

---

c) Exit of birds/eggs to other farms/establishments/hatcheries/abattoirs NO ❑ YES ❑

(From 21 days prior to the onset of the first clinical signs and the date the farm was put under restriction)

Date ..../..../.... N° .... Species ....

Destination: Other farm ❑ Hatchery ❑ Abattoir ❑ Other ....

Name of holding .... Code □□□ □□ □□□□

Address ....

CAP/Zip/Postal code .... Province/State/Country ....

---

d) Exit of birds/eggs to other fairs/markets/exhibitions NO ❑ YES ❑

(From 21 days before the onset of the first clinical signs and the date the farm was put under restriction)

Date ..../..../.... N° .... Species ....

Destination: Fair ❑ Market ❑ Exhibition ❑ Other ....

Address ....

CAP/Zip/Postal code .... Province/State/Country ....

2. Movement of people: possible means of introduction or of spread of infection

(From 21 days before the onset of the first clinical signs and the date the farm was put under restriction)
NO ❑ YES ❑

Date ...../...../..... Surname and first name ....

❑ Veterinarian ❑ Technician
❑ Vaccinating crew ❑ Debeaker
❑ Other farmer ❑ Dealer
❑ Other (specify) ....

Address ....

CAP/Zip/Postal code .... Province/State/Country ....

.... Phone number ....

Previously visited farm: Name ....

CAP/Zip/Postal code .... Province/State/Country ....

Date ....../....../......

---

3. Movement of mammals (only for avian influenza)

Introduction of mammals from other farms/markets/exhibitions NO ❑ YES ❑

(From 21 days before the onset of first clinical signs)

Date ...../...../..... N° .... Species .... Farm ❑ Market ❑ Exhibition ❑

Name of Farm..

Code ☐☐☐ ☐☐ ☐☐☐☐

Address..

CAP/Zip/Postal code .... Province/State/Country ....

Car plate of vehicle transporting animals ....

Movement of mammals to other destinations (e.g. farms/slaughterhouses) NO ❑ YES ❑

(From 21 days before the onset of the first clinical signs and the date the farm was put under restriction)

Date ...../...../..... N° .... Species ....

Farm ❑ Slaughterhouse ❑ Other ❑ ....

Name of company..

Code ☐☐☐ ☐☐ ☐☐☐☐

CAP/Zip/Postal code .... Province/State/Country ....

Car plate of vehicle transporting animals ....

Movement of Vehicles:

(A) Transport of animals, **(B)** Transport of feed, **(C)** Transport of eggs, **(D)** Collection of dead animals, **(E)** Fuel/Gas, **(Other)** Specify

(From 21 days before the onset of the first clinical signs and the date the farm was put under restriction)

Date of entry	Vehicle (A/B/C/D/E/ other)	Name of company	Fax/Phone number	Vehicle plate (tractor)	Vehicle plate (trailer)	Transporter (company name)	Driver	Phone number

a) Indirect contacts with other poultry establishments NO ❑ YES ❑

(Sharing of equipment, vehicles, feed, staff from 21 days before the onset of the first clinical signs and the date the farm was put under restriction)

Date of contact …../…../.....

Name of farm or holding .... Code □□□ □□ □□□□

Address ....

CAP/Zip/Postal code .... Province/State/Country ....

Species farmed .... Number ....

☐ Shared vehicle ☐ Shared feed

☐ Shared equipment ☐ Shared staff

☐ Collection/recycle of litter ☐ Other (specify) ....

---

b) Other farms owned by the same owner NO ❑ YES ❑

Name of farm or holding .... Code □□□ □□ □□□□

Address ....

CAP/Zip/Postal code .... Province/State/Country ....

Species farmed .... Number ....

Empty ❑ Full ❑

---

c) Poultry farms located near the outbreak NO ❑ YES ❑

Name of farm or holding .... Code □□□ □□ □□□□

Address ....

CAP/Zip/Postal code .... Province/State/Country ....

Distance in metres ....

Empty ❑ Full ❑

Anamnestic Data

Weekly mortality

NB: data concerning mortality rates recorded from 6 weeks prior to the onset of clinical signs

Dates		Number of dead animals
From	To	

NB: The weekly farm mortality record must be made available and signed by the farmer and official veterinarian.

Remarks: ....
....
....

Date of onset of clinical signs of avian influenza ..../........./.........

Clinical signs observed by the farmer: ....
....
....
....

Total number of birds (dead or alive)	Number of ill birds (at the time of implementation of restriction measures)	Number of dead birds (at the time of implementation of restriction measures)	Number of birds depopulated

NB: this information must refer to the data collected when the farm was put under restriction, with mortality and morbidity related to the suspicion of avian influenza only.

Vaccination Programme

Vaccination of birds is practised NO ❑ YES ❑

Date of vaccination	Type of vaccine[a]	Commercial name	Administration route

[a]Live or inactivated

Vaccinating Staff:

❑ Family ❑ Employees ❑ External staff ❑ Other ....

Remarks ....

Administration of Drugs/Medications

In the last 15 days: NO ❑ YES ❑ (specify) ....

Starting day	End of administration	Type of drug	Commercial name	Administration route

Staff who Administered the Medication:

❑ Family ❑ Employees ❑ External staff ❑ Other ....

Remarks ....

Clinical and gross findings: registration form

Clinical investigation form (to be filled in for each species affected)

SPECIES ....

Weights
- on target YES ❑ NO ❑
- below target YES ❑ NO ❑
- above target YES ❑ NO ❑

Feed consumption
Increased ❑ Decreased ❑

Water consumption
Increased ❑ Decreased ❑

Depression ❑
Respiratory signs: mild ❑
severe ❑
Drop or cessation of egg production ❑
Oedema, cyanosis or cutaneous haemorrhages ❑
Diarrhoea ❑
Nervous signs ❑
Sudden death ❑
Individual ❑
Mass ❑
Other ....

Flock appearance

Mortality rates .... % In the last day .... % In the last week .... %

Individual sudden death ❑ Sudden mass mortality ❑ .... %

Reduction of feed consumption ❑ .... % Reduction of water consumption ❑

Reduction in body weight ❑ .... %

Ruffled feathers ❑ Huddled ❑ Anorexia ❑ Depression ❑ Thirst ❑

Egg production

Drop .... % Cessation ❑
Change in colour ❑
Change in quality ❑
Date of onset .... Duration ....

Respiratory symptoms

Coughing ❑ Oedema of head/neck ❑ Gasping ❑ Conjunctivitis ❑
Sneezing ❑ Modified vocalisation ❑ Rales ❑ Sinusitis ❑
Nasal discharge ❑ Head shaking ❑

Enteric signs

Diarrhoea – haemorragic ❑ – white ❑ – frothy ❑
– green ❑ – foul smelling ❑

Nervous signs

Tremors	❑	Ataxia	❑	Paresis	❑
Circling movements	❑	Blindness	❑	Incoordination	❑
Recumbency	❑	Paralysis	❑	Torticollis	❑

Locomotive disorders

Dropped wings ❑ Abnormal gait ❑

Detail: ....

Vascular/cutaneous disorders

Oedema ❑ (indicate where) .... Cyanosis ❑ ....

Haemorrhages ❑ .... Pallor ❑ ....

Other:

DISEASE SUSPECTED: ❑ Newcastle disease ❑ Avian influenza

Gross finding investigation form (to be filled in for each species affected)

Rhinitis and sinusitis		❑
Tracheitis	*catarrhal*	❑
	haemorrhagic	❑
Airsacculitis		❑
Polmonitis		❑
Enteritis	*catarrhal*	❑
	haemorrhagic	❑
Pancreatitis		❑
Haemorrhages	*epicardium*	❑
	endocardium	❑
	proventriculus	❑
	ovarian follicles	❑
	kidney	❑
	liver	❑
	proventriculus	❑
	cecal tonsil	❑
	Peyer's plaques	❑
	muscles	❑
Necrosis	*kidney*	❑
	liver	❑
	pancreas	❑
	spleen	❑
General condition	*good*	❑
	Below average	❑
	poor	❑

Other ....

Remarks ....

....

....

Signature

....

Check:
- Daily mortality record
- Egg production record
- Feed consumption record
- Data on average weight gain
- Information on water consumption
- Record of movements to and from farm

Annex 3
Biosafety Procedures

William G. Dundon

Introduction

Avian influenza (AI) is a highly contagious disease of birds and, although it shares similarities with human influenza, the viruses responsible for AI cannot be transmitted easily to humans. However, AI viruses have been transmitted to humans sporadically and under specific conditions that have included:

- Direct contact with sick or dead birds
- Contact with surfaces that have been contaminated with the excreta (faeces) and/or secretions of infected birds
- Contact of oral, ocular or nasal mucosa with infected aerosols
- Consumption of uncooked meat or blood of infected birds.

Accordingly, the issue of biosafety amongst personnel who are in contact with the virus is of the utmost importance. Personnel at risk of exposure to the virus fall into two categories. The first group consists of those who are involved in the control of outbreaks and the eradication of AI and whose tasks include the culling of infected birds, carcase elimination and the cleaning and disinfection of premises. The second group at risk of exposure includes laboratory-based personnel who are involved in the reception and processing of contaminated specimens and samples that contain the virus. The recommended biosafety precautions for both groups are listed below.

Group 1 or Field-Based Personnel

Recommendations

- The number of people involved in depopulation/stamping-out operations must be kept to a minimum.
- The quicker and more efficient the control of an AI outbreak in an infected flock, the lower the risk of spread to personnel.
- It is essential that personnel are aware of, and adopt, adequate biosafety precautions that reduce the risk of further spread of the virus and of contamination of themselves and others.
- Personnel should be informed that smoking and eating are forbidden in the work area and that the touching of the nose, mouth and eyes with potentially contaminated hands should be avoided.
- At the end of a depopulation/stamping-out operation, all of the personal protective equipment must be either suitably disposed of or adequately cleaned and disinfected.

Personal Protective Equipment

Personnel who are at risk of infection or who are in direct contact with potentially infected birds must wear the following personal protective equipment:

- Protective, disposable, long-sleeved overalls and an impermeable apron.
- Disposable head covers/bouffant caps.
- Rubber or polyurethane boots that can be washed and disinfected, or disposable shoe-covers.
- Rubber gloves that can be washed and disinfected; disposable nitrile (synthetic latex) gloves. Cotton gloves may be worn as under-gloves to prevent irritation of the hands due to prolonged use of synthetic over-gloves. Gloves should be replaced immediately if they show signs of wear and deterioration. The gloves should be removed with care and without touching non-contaminated surfaces.
- Disposable protective FFP2D face mask (or equivalent) that fits the face correctly.

I. Capua, D.J. Alexander (eds.) *Avian Influenza and Newcastle Disease*,

- Protective visor to protect against splashes from contaminated material.
- Protective goggles.

After use, the personal protective equipment should be removed and the hands washed and disinfected in the following order:

1. Gloves
2. Overalls
3. Washing and disinfection of hands
4. Protective goggles
5. Visor and face mask
6. Washing and disinfection of hands.

In steps 3 and 6, following removal of the personal protective equipment, personnel should wash their hands with soap and water for at least 15–20 s.

Health Considerations

- When possible, unvaccinated personnel should be vaccinated with the current season's influenza virus to reduce the possibility of dual infection and reassortment between an avian and a human virus.
- Personnel should receive a suitable antiviral drug daily for the duration of direct contact with infected poultry or their secretions and for 5–7 days after the last day of potential exposure to the virus.
- Personnel should be told to monitor their health and report any clinical symptoms that include fever, respiratory difficulties and/or conjunctivitis for 1 week after potential exposure to the virus.

Group 2 or Laboratory-Based Personnel

- Laboratory safety is the responsibility of all supervisors and laboratory employees. Individual workers are responsible for their own safety and that of their colleagues.
- Good laboratory practise is fundamental to laboratory safety and varies depending on the risk associated with the infectious material being manipulated.

Animal clinical specimens from suspect highly pathogenic avian influenza (HPAI) cases may be tested by polymerase chain reaction (PCR) assays using standard biosafety level 2 (BSL 2) work practises in a Class II biological safety cabinet. Commercial antigen detection testing can also be conducted under BSL 2 levels to test for influenza. In addition, specimens may be processed for packaging and distribution to diagnostic laboratories for further testing in a BSL 2 laboratory.

HPAI viruses must be worked with under biosafety level 3+ laboratory conditions. Therefore, manipulations involving growth of the agent should be in a BSL 3+ laboratory using BSL 3+ operational practises. These include controlled-access double-door entry with changing room and shower, use of respirators, decontamination of all wastes and showering of all personnel (see below).

NB: When laboratories do not meet BSL 2 containment conditions, specimens from suspected or confirmed HPAI cases should be sent to suitably equipped reference laboratories for further processing

Below is a list of guidelines that must be taken into considerations when working in BSL 2 and BSL 3 laboratories. These guidelines were adapted from Biosafety in Microbiological and Biomedical Laboratories, 5th edn., Feb 2007 (http://www.cdc.gov/OD/ohs/biosfty/bmbl5/bmbl5toc.htm).

Guidelines for Working in BSL 2 Laboratories

- Biosafety level 2 is applicable for work involving agents of moderate potential hazard to personnel and the environment.
- Laboratory personnel must have specific training in handling pathogenic agents.
- Access to the laboratory must be limited when work is being conducted.
- Extreme precautions must be taken with contaminated sharp items, including needles and syringes, slides, pipettes, capillary tubes and scalpels. Broken glassware must not be handled directly by hand, but must be removed by mechanical means such as a brush and dustpan, tongs or forceps. Containers of contaminated needles, sharp equipment and broken glass must be decontaminated before disposal.
- Procedures in which infectious aerosols or splashes should be minimised must be conducted in biological safety cabinets.
- Personnel must wash their hands after handling viable materials, after removing gloves and before leaving the laboratory.
- Eating, drinking, smoking, handling contact lenses and applying cosmetics are not permitted in the work areas.

- Mouth pipetting must be expressly forbidden.
- Work surfaces must be decontaminated on completion of work and immediately after any spill or splash of viable material. Disinfectants that are effective against the agents of concern must be used.
- All cultures, stocks and other regulated wastes must be decontaminated before disposal by an approved decontamination method such as autoclaving. Materials to be decontaminated outside of the immediate laboratory must be placed in a durable, leak-proof container and closed for transport from the laboratory.
- A biohazard sign must be posted on the entrance to the laboratory when aetiologic agents are in use. Appropriate information must include the agent(s) in use, the biosafety level, the required immunisations, the investigator's name and telephone number, any personal protective equipment that must be worn in the laboratory and any procedures required for exiting the laboratory.
- A biosafety manual should be prepared specifically for the laboratory by the laboratory director. Personnel should be advised of special hazards and must read and follow instructions on practices and procedures.
- The laboratory director must ensure that laboratory personnel receive appropriate training on the potential hazards associated with the work involved and the necessary precautions to prevent exposures.
- Cultures, tissues, specimens of body fluids or potentially infectious wastes must be placed in a container with a cover that prevents leakage during collection, handling, processing, storage, transport or shipping.
- Contaminated equipment must be decontaminated accordingly before it is sent for repair or maintenance or packaged for transport.
- Spills and accidents that result in overt exposures to infectious materials are immediately reported to the laboratory director. Medical evaluation, surveillance and treatment are provided as appropriate and written records are maintained.
- Properly maintained biological safety cabinets, preferably Class II, must be used whenever:
  - Procedures with a potential for creating infectious aerosols or splashes are conducted, e.g. centrifuging, grinding, blending, vigorous shaking or mixing, sonic disruption, opening containers of infectious materials whose internal pressures may be different from ambient pressures and harvesting infected tissues from animals or embryonated eggs.
  - High concentrations or large volumes of infectious agents are used. Such materials may be centrifuged in the open laboratory if sealed rotor heads or centrifuge safety cups are used, and if these rotors or safety cups are opened only in a biological safety cabinet.
- Face protection (goggles, mask, face shield or other splatter guard) must be used when the infectious agent is manipulated outside the biological safety cabinet.
- Protective laboratory coats, gowns, smocks or uniforms designated for laboratory use must be worn while in the laboratory. This protective clothing must be removed and left in the laboratory before leaving for non-laboratory areas. All protective clothing is either disposed of in the laboratory or laundered by the institution.
- Gloves must be worn when hands can come into contact with potentially infectious materials, contaminated surfaces or equipment. Wearing two pairs of gloves may be appropriate. Gloves must be disposed of when contaminated and removed when work with infectious materials is completed or when the integrity of the glove is compromised. Disposable gloves are not washed or reused. "Clean" surfaces should not be touched by gloved hands (e.g. keyboards, telephones). Hands should be washed following removal of gloves.

Structure of a BSL 2 Laboratory

- Each laboratory must contain a sink for hand washing that is "hands free" or preferably operated automatically.
- The laboratory should be designed so that it can be easily cleaned.
- Bench tops should be impervious to water and resistant to both moderate heat and the solutions used to decontaminate the work surfaces and equipment.
- Spaces between benches, cabinets and equipment should be accessible for cleaning. Chairs and other furniture used in laboratory work should be covered with a non-fabric material that can be easily decontaminated.
- Biological safety cabinets should be installed so that fluctuations of the room's air supply and of the exhaust air do not cause the biological safety cabinets to operate outside their parameters for

containment. Biological safety cabinets should be located away from doors and windows that can be opened in order to maintain the cabinets' airflow parameters for containment.

- An eyewash station should be readily available.
- Illumination must be adequate for all activities and should avoid reflections and glare that could impede vision.
- If the laboratory has windows that open to the exterior, they should be fitted with fly screens.

Guidelines for Working in BSL 3 Laboratories

Biosafety level 3 is applicable when work is done with agents that may cause serious or potentially lethal disease as a result of exposure by the inhalation route. Laboratory personnel must have specific training in handling pathogenic and potentially lethal agents.

- All procedures involving the manipulation of infectious materials must be conducted within biological safety cabinets or by personnel wearing appropriate personal protective clothing and equipment.
- The exhaust air from the laboratory room must be discharged to the outdoors through high-efficiency particulate air (HEPA) filters.
- Ventilation to the laboratory must be balanced to provide directional airflow into the room.
- Access to the laboratory must be restricted when work is in progress.
- Personnel must wash their hands after handling infectious materials, after removing gloves and when leaving the laboratory.
- Eating, drinking, smoking, handling contact lenses and applying cosmetics is forbidden in the laboratory. Persons who wear contact lenses in laboratories should also wear goggles or a face shield.
- Mouth pipetting must be strictly forbidden.
- Extreme precautions must be taken with contaminated sharp items including needles and syringes, slides, pipettes, capillary tubes and scalpels. Broken glassware must not be handled directly by hand, but must be removed by mechanical means such as a brush and dustpan, tongs or forceps. Containers of contaminated needles, sharp equipment and broken glass must be decontaminated before disposal. Plasticware should be substituted for glassware whenever possible.
- All procedures must be performed carefully to minimise the creation of aerosols.
- Work surfaces should be immediately decontaminated after any spill of contaminated material and at the end of the working day.
- All cultures, stocks and other regulated wastes must be decontaminated before disposal by an approved decontamination method, such as autoclaving. Materials to be decontaminated outside of the immediate laboratory must be placed in a durable, leak-proof container and closed for transport from the laboratory. Infectious waste from BSL 3 laboratories should be decontaminated before removal for off-site disposal.
- Laboratory doors must be kept closed when experiments are in progress.
- When infectious materials or infected animals are present in the laboratory, a hazard warning sign, incorporating the universal biohazard symbol, must be posted on all laboratory and animal-room access doors. The hazard warning sign must identify the agent, list the name and telephone number of the laboratory director or other responsible personnel and indicate any special requirements for entering the laboratory, such as the need for immunisations, respirators or other personal protective measures.
- A biosafety manual specific to the laboratory must be prepared by the laboratory director. Personnel must be advised of special hazards and required to read and follow instructions on practices and procedures.
- Laboratory and support personnel must receive appropriate training on the potential hazards associated with the work involved and the necessary precautions to prevent exposures.
- The laboratory director is responsible for ensuring that, before working with organisms at BSL 3, all personnel demonstrate proficiency in standard microbiological practices and techniques and in the practices and operations specific to the laboratory facility.
- All open manipulations involving infectious materials must be conducted in biological safety cabinets. Work in open vessels must not be conducted on the open bench.
- Laboratory equipment and work surfaces should be decontaminated routinely with an effective disinfectant, after work with infectious materials is finished and especially after spills, splashes or other contamination with infectious materials.

- Contaminated equipment must be decontaminated before removal from the facility for repair or maintenance or packaging for transport.
- Cultures, tissues, specimens of body fluids or wastes must be placed in a container that prevents leakage during collection, handling, processing, storage, transport or shipping.
- All potentially contaminated waste materials (e.g. gloves, lab coats) from laboratories must be decontaminated before disposal or reuse.
- Spills and accidents that result in overt or potential exposures to infectious materials must be immediately reported to the laboratory director. Appropriate medical evaluation, surveillance and treatment should be provided and written records must be maintained.
- Protective laboratory clothing such as solid-front or wrap-around gowns, scrub suits, or coveralls must be worn by personnel when in the laboratory. Protective clothing must not be worn outside the laboratory. Reusable clothing must be decontaminated before being laundered.
- Gloves must be worn when handling infectious materials, infected animals and contaminated equipment. Frequent changing of gloves accompanied by hand washing is recommended. Disposable gloves must not be reused.
- All manipulations of infectious materials, necropsy of infected animals, harvesting of tissues or fluids from infected animals or embryonated eggs , etc., must be conducted in a Class II or Class III biological safety cabinet.

Structure of a BSL 3 Laboratory

- Access to the laboratory must be restricted. Passage through a series of two self-closing doors is the basic requirement for entry into the laboratory from access corridors. Doors must be lockable.
- Each laboratory room must contain a sink for hand washing. The sink should be hands-free or automatically operated and located near the room exit door.
- The interior surfaces of walls, floors and ceilings of areas where BSL 3 agents are handled should be constructed for easy cleaning and decontamination. Seams, if present, must be sealed. Walls, ceilings and floors should be smooth, impermeable to liquids and resistant to the chemicals and disinfectants normally used in the laboratory.
- All windows in the laboratory must be closed and sealed.
- A method for decontaminating all laboratory wastes must be available in the facility and utilised, preferably within the laboratory (i.e. autoclave, chemical disinfection, incineration).
- A system that creates directional airflow that draws air into the laboratory from "clean" areas and toward "contaminated" areas must be used. The exhaust air must not be recirculated to any other area of the building. The outside exhaust must be dispersed away from occupied areas and air intakes must be HEPA-filtered.
- Continuous-flow centrifuges or other equipment that may produce aerosols must be contained in devices that exhaust air through HEPA filters before discharge into the laboratory.
- An eyewash station must be readily available inside the laboratory.
- Illumination is adequate for all activities, avoiding reflections and glare that could impede vision.

Decontamination of Working Surfaces

- A general-purpose laboratory disinfectant should have a concentration of 1 g chlorine/l (0.1%), e.g. domestic bleach diluted 1:50.
- A stronger laboratory disinfectant should have a concentration of 5 g chlorine/l (0.5%), e.g. domestic bleach diluted 1:10. This should be used for the disinfection of biohazardous spills and spills that contain large amounts of organic material.
- Bleach solutions should be made up freshly and allowed a contact time of at least 10 min.

Annex 4
Laboratory Solutions

William G. Dundon

Phosphate-Buffered Saline (PBS) Solution

For 1 litre

1. Mix:
   - *0.2 g* KH_2PO_4
   - *2.9 g* $Na_2HPO_4 + 2H_2O$*:*
   - *0.2 g KCl*
   - *8 g NaCl*
2. Adjust pH to 7.2–7.4 with HCl.
3. Add distilled water to 1000 ml.
4. Sterilise in autoclave at 121°C for 20 min.
5. The solution can be stored at +4°C for a maximum of 1 year.

Phosphate-Buffered Saline (PBS) Solution + Bovine Albumin (0.05%)

1. Add 0.25 g bovine albumin per 500 ml PBS solution.
2. The solution can be stored at +4°C for a maximum of 1 week.

PBS + Antibiotics

To PBS add:

- *10000 IU Penicillin/ml*
- *10 mg Streptomycin/ml*
- *5000 IU Nystatin/ml*
- *250 µg Gentamycin sulphate/ml*

Reagents for the Neuraminidase Inhibition (NI) Test

Standard Fetuin Substrate

For 1 litre

1. To 48 g fetuin add distilled water to 1000 ml.
2. Dilute this solution 1:2 with PBS (pH 5.9) containing 6 mM $CaCl_2.6\ H_2O$ before use.

0.025 M Sodium Periodate in 0.125 N Sulphuric Acid

For 500 ml

1. Add 2.67 g sodium periodate to 500 ml of a 0.125 N H_2SO_4 solution.
2. Mix well.

2% (w/v) Sodium Arsenite in 0.5 N HCl

For 100 ml

1. Add 2 g sodium arsenite ($NaAsO_2$) to 100 ml of 0.5 N HCl.
2. Mix well.

0.1M Thiobarbituric Acid (TBA)

1 litre

1. Add 7.2 g TBA to 400 ml distilled water.
2. Adjust to pH 9.0 with NAOH.
3. Add distilled water to 500 ml.

I. Capua, D.J. Alexander (eds.) *Avian Influenza and Newcastle Disease*,

5% (v/v) 10N HCl Butanol Acid

1. Add 5 ml concentrated hydrochloric acid to 100 ml butane-1-ol.
2. Mix well.

ALSEVER Solution

- *20.5 g Glucose*
- *8 g $C_6H_5Na_3O_7$ (sodium citrate)*
- *0.55 g $C_6H_8O_7$ (citric acid)*
- *4.2 g NaCl*

1. Add distilled water to 1000 ml.
2. Sterilise with 0.45-µm filters.
3. The solution can be stored at +4° C for a maximum of 6 months.

Agarose Gels

Gel Buffer–TAE 10× (use 1×)

1 litre

- *484 g Trizma base*
- *11.4 ml Glacial acetic acid*
- *20 ml EDTA 0.5M (pH 8)*

1. Add distilled water to 1000 ml.
2. Autoclave at 121°C for 20 min.
3. The solution can be stored at +4°C for a maximum of 1 year.

Gel-Loading Buffer 10× (use 1×)

10 ml solution

- *1 ml TAE 10×*
- *6 ml Glycerol (100%)*
- *1 ml Bromophenol blue and xylene cyanol (0.5% w/v)*

1. Mix the components and add distilled water to 10 ml.
2. The buffer can be stored at +4°C for a maximum of 1 year.

Annex 5
Guidelines for Shipping Avian Influenza and Newcastle Disease Virus Samples to OIE Reference Laboratories

William G. Dundon

Below is a list of the seven World Organisation of Animal Health (OIE) Reference Laboratories for avian influenza and Newcastle disease including contact details. Diagnostic samples (isolated viruses, clinical specimens and swabs) can be sent to any of these laboratories in compliance with the guidelines listed below.

NB: The reference laboratories **must** be contacted prior to the shipping of samples in order to procure the necessary importation documents and to expedite clearance through customs.

Dr. Ian Brown
VLA Weybridge
New Haw, Addlestone, Surrey KT15 3NB
United Kingdom
Tel: (44.1932) 34.11.11 Fax: (44.1932) 34.70.46
Email: i.h.brown@vla.defra.gsi.gov.uk

Dr. Ilaria Capua
Istituto Zooprofilattico Sperimentale delle Venezie, Laboratorio di Virologia
Viale dell'Università, 10, 35020 Legnaro, Padova
Italy
Tel: (39.049) 808.43.69 Fax: (39.049) 808.43.60
Email: icapua@izsvenezie.it

Dr. Timm C. Harder
Federal Research Centre for Virus Diseases of Animals (BFAV), Institute of Diagnostic Virology
Boddenblick 5a, 17493 Greifswald - Insel Riems
Germany
Tel: (49.383) 51.71.96 Fax: (49.383) 51.72.75
Email: timm.harder@fli.bund.de

Dr. Hiroshi Kida
Graduate School of Veterinary Medicine, Hokkaido University, Department of Disease Control
Kita-18, Nishi-9, Kita-ku, Sapporo 060-0818
Japan
Tel: (81.11) 706.52.07 Fax: (81.11) 706.52.73
Email: kida@vetmed.hokudai.ac.jp

Dr. Brundaban Panigrahy
National Veterinary Services Laboratories
P.O. Box 844, Ames, IA 50010
United States of America
Tel: (1.515) 663.75.51 Fax: (1.515) 663.73.48
Email: brundaban.panigrahy@aphis.usda.gov

Dr. John Pasick
Canadian Food Inspection Agency, National Centre for Foreign Animal Disease
1015 Arlington Street, Winnipeg, Manitoba R3E 3M4
Canada
Tel: (1.204) 789.20.13 Fax: (1.204) 789.20.38
Email: jpasick@inspection.gc.ca

Dr. Paul W. Selleck
CSIRO, Australian Animal Health Laboratory (AAHL)
5 Portarlington Road, Private Bag 24, Geelong 3220, Victoria
Australia
Tel: (61.3) 52.27.50.00 Fax: (61.3) 52.27.55.55
Email: paul.selleck@csiro.au

Types of Specimens

Specimens submitted to the OIE Reference laboratories may be virus isolates or clinical specimens, such as tissues, swabs, blood or serum, collected from birds that are known or suspected to be infected by avian influenza or Newcastle disease virus.

The packing of specimens and their shipment to external laboratories by air is complex and is governed by international and national regulations.

I. Capua, D.J. Alexander (eds.) *Avian Influenza and Newcastle Disease*,

International air transport of specimens known or suspected to contain the avian influenza or Newcastle disease agent **must** follow the guidelines in the current edition of the International Air Transport Association's (IATA) Dangerous Goods Regulations (*Infectious Substances Shipping Guidelines* 2007).

Categories covering the shipment of specimens by air

- **Category A** refers to infectious substances that are capable of causing permanent disability or life-threatening or fatal disease in otherwise healthy humans or animals. Highly pathogenic avian influenza (HPAI) virus falls under this category when shipped as "cultures only". Thus, cultures of HPAI (i.e. virus isolates) must be transported as **Category A** and must be assigned the number **UN 2814** (infectious substance affecting humans and animals). Avian paramyxovirus type 1 (i.e. velogenic Newcastle disease virus) "cultures only" are also included under **Category A** and must be assigned the number **UN 2900** (infectious substance affecting animals).
- **Category B** refers to all other infectious substances that are not included in Category A. Regarding the shipment of animal samples suspected or confirmed to contain HPAI viruses, animal blood and other animal samples (tissue, swabs, faeces) known or suspected to contain this virus can be transported as "diagnostic specimens" (**UN 3373**) and are included in Category B. Likewise, animal samples suspected or confirmed to contain velogenic Newcastle disease virus can be transported as "diagnostic specimens" (**UN 3373**).

NB: Individual airlines may have their own policies and these may be stricter than those issued by IATA.

- **Packaging Requirements: Category A**
  Triple packaging system: Infectious substances of Category A may only be transported in packaging that meets UN class 6.2 specifications and packing instruction P620 (PI602) as described below. Packaging should be composed of:
  1. Primary receptacle: *A labelled, primary, watertight, leak-proof receptacle containing the specimen. The receptacle should be wrapped in enough absorbent material to absorb all fluids in case of breakage.*
  2. Secondary receptacle: *A second durable, watertight, leak-proof receptacle to enclose and protect the primary receptacle(s). Several wrapped primary receptacles may be placed in one secondary receptacle. Sufficient additional absorbent material must be used to cushion multiple primary receptacles.*
  3. Rigid outer packaging: *The secondary receptacle is placed in an outer shipping package that protects it and its contents from outside influences, such as physical damage and water, while in transit. This packaging must bear the UN packaging specification label **UN 2814** (for avian influenza) or **UN 2900** (for Newcastle disease virus).*

Labelling of the outer package for shipment of infectious substances **must** include the following elements:

a. The International Infectious Substance Label
b. Address label containing:
   - The receiver's name, address and telephone number
   - The shipper's name, address and telephone number
   - The UN shipping name (e.g. Infectious Substance Affecting Humans and/or Animals) followed by the scientific name (e.g. avian influenza virus, Newcastle disease virus)
   - The UN number (e.g. UN2814 or UN2900)

NB: If the outer package is further packed in an overpack (i.e. polystyrene container with dry ice) both the outerpack and the overpack must carry the above information and the overpack must have a label stating "*Inner packages comply with prescribed specifications*".

- **Required shipping documents: Category A**
- *Shipper's Declaration of Dangerous Goods*
- *A detailed packaging list that includes the receiver's address, number of the packages, details of contents, weight and value*
- *Airway bill (if shipping by air)*
- *An import and/or export permit and/or declaration, if required*

NB: For import permits, please contact the OIE Reference Laboratory. For export permits, please follow your own country's guidelines.

NB: If the outer package contains primary receptacles exceeding 50 ml, two *Orientation Labels* indicating the "*UP*" direction with an arrow must be placed on opposite sides of the package.

NB: Specimen data forms, letters and other types

of information that identify or describe the specimen and also identify the shipper and receiver should be taped to the outside of the secondary receptacle.

NB: The maximum net quantity of infectious substances that can be contained in a package is 50 ml or 50 g if it is transported by passenger aircraft and 4 l or 4 kg if it is transported by cargo aircraft.

NB: International air carriers strictly prohibit the onboard, hand carriage of infectious substances.

- **Packaging Requirements: Category B**

– **Triple packaging system:** Infectious substances of Category B may only be transported in packaging that meets UN class specifications and packing instruction P650 as described below. The packaging shall be of good quality, strong enough to withstand the shocks and loadings normally encountered during transport. Packaging shall be constructed and closed in such a way as to prevent any loss of contents that might be caused under normal conditions of transport by vibration or by changes in temperature, humidity or pressure.

1. Primary receptacle: *A labelled, primary, watertight, leak-proof receptacle containing the specimen. The receptacle should be wrapped in enough absorbent material to absorb all fluids in case of breakage.*
2. Secondary receptacle: *A second durable, watertight, leak-proof receptacle to enclose and protect the primary receptacle(s). Several wrapped primary receptacles may be placed in one secondary receptacle. Sufficient additional absorbent material must be used to cushion multiple primary receptacles.*
3. Rigid outer packaging: *The secondary receptacle is placed in an outer shipping package that protects it and its contents from outside influences, such as physical damage and water, while in transit. This packaging must bear the UN packaging Specification Markings **UN3373**.*

NB: For air transport, no primary receptacle shall exceed 1 l. Also, the volume shipped per package shall not exceed 4 l or 4 kg.

- **Required shipping documents: Category B**

A shipper's Declaration of Dangerous Goods declaration is **not** required for category B infectious substances. The following documents are required:

- *A detailed packaging list that includes the receiver's address, number of the packages, details of contents, weight and value*
- *Airway bill (if shipping by air)*
- *An import and/or export permit and/or declaration, if required*

NB: For import permits please contact the OIE Reference Laboratory. For export permits please follow your own country's guidelines.

References

For more information on shipping infectious agents, please refer to the following sites:

www.iata.org

www.who.int/csr/emc97_3.pdf

www.who.int/csr/resources/publications/biosafety/WHO_CDS_EPR_2007_2/en/index.html

http://www.unece.org/trans/danger/publi/unrec/rev14/14files_e.html

Flowchart for the classification of avian influenza and Newcastle disease specimens

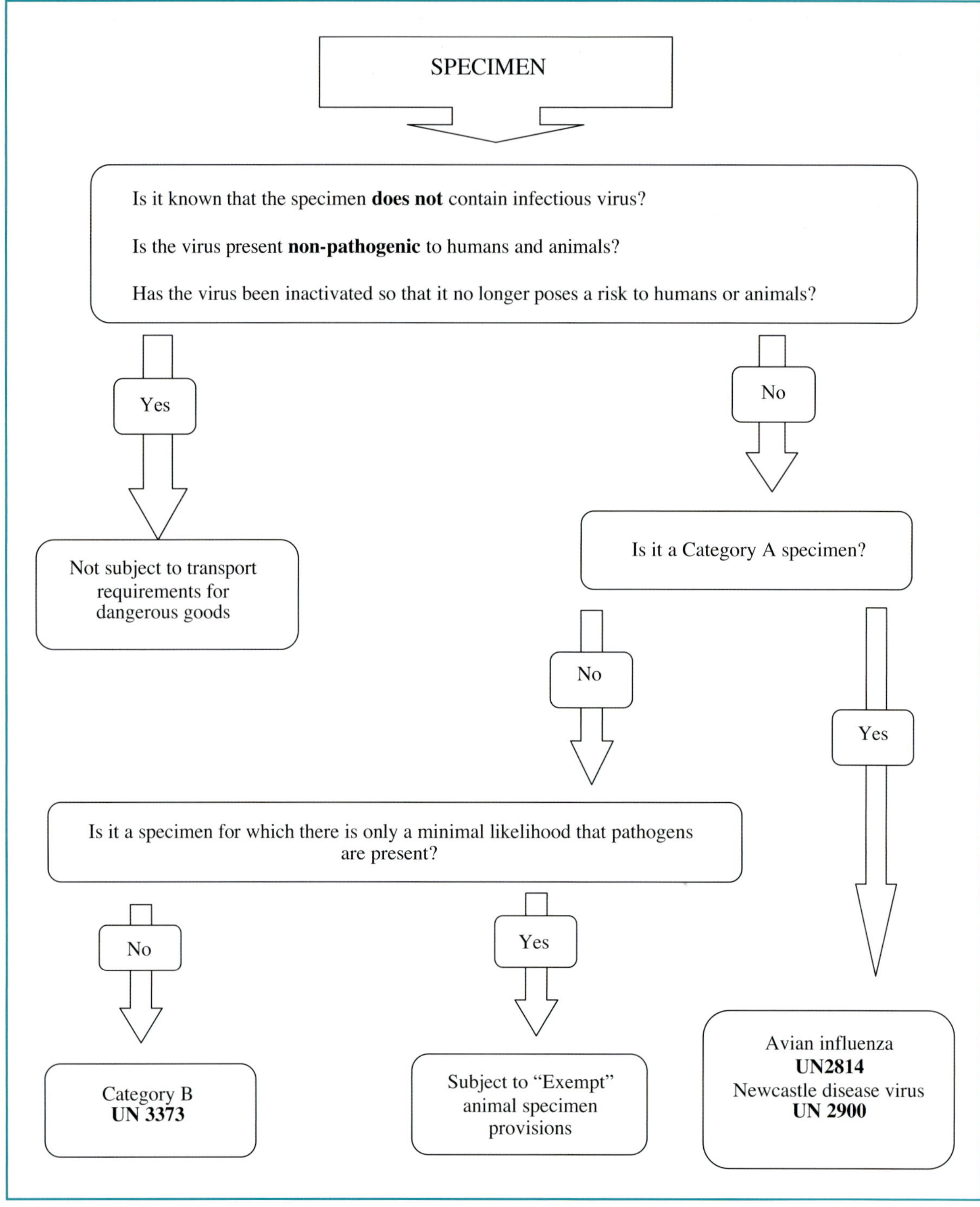

Guideline Summary Table

Sample type	Shipping temperature	Antibiotic medium[a]	UN code for avian influenza	UN code for Newcastle disease virus
Isolated virus	Dry ice	No	Infectious substance UN 2814	Infectious substance UN 2900
Organ (large)[b]	Dry ice	No	Diagnostic specimen UN 3373	Diagnostic specimen UN 3373
Organ (small)[c]	Dry ice	Yes	Diagnostic specimen UN 3373	Diagnostic specimen UN 3373
Faeces[c]	Dry ice	Yes	Diagnostic specimen UN 3373	Diagnostic specimen UN 3373
Swab[d]	Dry ice	Yes	Diagnostic specimen UN 3373	Diagnostic specimen UN 3373
Blood/serum	4°C	No	Diagnostic specimen UN 3373	Diagnostic specimen UN 3373

[a] Antibiotic medium: 10,000 units penicillin/ml, 10 mg streptomycin/ml, 0.25 mg gentamycin/ml, 5,000 units mycostatin/ml It is imperative when making the medium that the pH is checked after the addition of antibiotics and re-adjusted to pH 7.0–7.4 before use.

[b] Trachea, lungs, duodenum, caecal tonsil, brain tissue, liver, spleen and other obviously affected organs from recently dead birds.

[c] Faecal samples and small pieces of organs should be homogenized in antibiotic medium[a] to give a suspension of 10–20% (w/v). The suspensions should be left for about 2 h at ambient temperature (or longer periods at 4°C) and then clarified by centrifugation (e.g. 1000 × g for 10 min).

[d] Swabs should be placed in sufficient antibiotic solution to ensure complete immersion.

Subject Index

A

Acute fowl cholera 70
Aetiology 1, 19
Agar gel immunodiffusion (AGID) assay 81
Agarose gel 96-106, 128-130
Ambion's protocol 89
Amplicons 96, 97, 103
Amplicons detection and analysis 96, 97, 103
Antibodies 2, 6, 13, 20, 23, 24, 29, 41, 80, 81, 83, 85, 113, 124
APMV-1 19, 20, 22, 113, 116, 117, 124, 127-131
Avian Influenza (AI)
– characterisation 7, 14, 30, 76, 87
– clinical signs 2, 4-7, 10, 13
– conventional diagnosis 73
– epidemiology 1, 6, 8, 13
– farm visit 32, 45, 67
– guidelines for shipping 177
– H5 influenza virus 1-14, 29, 45, 87, 103, 108, 109
– H7 influenza virus 1, 4, 6, 8, 29, 45, 79, 87, 105, 110
– highly pathogenic avian influenza (HPAI) 1-11, 29, 43, 45, 52-69, 75, 80, 87, 103, 105, 108-110, 151, 170, 178
  differential diagnosis 45, 68, 69
– isolation 73
– low pathogenic avian influenza (LPAI) 2-9, 29, 45-52, 68, 69, 103, 105, 108, 110
– low pathogenicity 5, 7, 28, 41, 45, 46
– molecular diagnosis 20, 87
– notification 9, 13, 27-29, 31, 32
– outbreak 1-5, 7-11, 13, 31, 133, 138
– pathology 7, 45
– protocol
  H5 103
  H7 105
  IZSVe protocol 89
  real-time 109
– serological test 80, 81
– virus 1, 5, 11, 12, 45, 73, 76, 78, 79, 83, 103, 105
  characterisation 30, 76, 87, 123, 124
  Type A 1, 78, 80, 81, 87, 101, 107
– virus isolation 73
  HA test 76
  methods 73

B

Bin composting 143
Biosafety 35, 169-172, 179
Blood 39, 41, 66, 67, 70, 76, 78, 83, 88, 95, 169, 177, 178, 181
Burial 34, 139-141, 147, 149

C

Caged birds 5, 10, 21, 147
Carcases 24, 34, 39, 43, 117, 134, 139-149
Cell counting chamber 83
Cell cultures 123
Chickens 2, 4, 5, 7, 9, 13-26, 29, 30, 45, 49, 53, 55, 61, 69, 70, 80, 83, 85, 113, 114, 116, 117, 119-121, 124, 125
Cleavage 1, 2, 4, 16-18, 22, 25, 26, 29, 103, 105
Cleavage-site sequence 2
Clinical examination 32, 68
– signs 2, 4-7, 10, 13, 20, 24, 28-30, 32, 33, 39, 45, 49, 51-55, 59, 61, 63, 68-70, 80, 113-117, 119-121, 124, 125, 160-164
Cloacal swabs 23, 39, 42, 73, 94
Collection of specimens 39
Compartment 28, 29
Composting 142-145, 147-150
Cormorants 20, 21, 25, 26

D

Decontamination 133-139, 141, 143, 145-147, 149, 170-173

– formaldehyde gas 146
Destruction of animals 146, 149
Differential diagnosis 45, 47, 49, 51, 53, 55, 57, 59, 61, 63, 65, 67-69, 71, 113, 115, 117, 119, 121
Disinfectant 34, 35, 133-138, 153, 172, 173
Disinfection unit 32
Disposal of carcases 34, 139, 141
Distribution 8, 23, 32, 95, 134, 137, 145, 170
Dogs 7
Domestic poultry 5, 7, 9, 20, 68
Duck virus enteritis (Duck plague) 70
Ducks 2, 4, 5, 8-11, 14-18, 20, 52, 63, 70, 116, 122, 147

E
Education 13
ELISA 81, 83
Embryonated eggs 73, 75, 123, 171, 173
End-point RT-PCR 96, 101, 103, 105, 127, 129
Endonuclease analysis 127, 128
Epidemiological enquiry 33
Epidemiology 1, 6, 8, 13, 19, 20, 28, 32, 139
Ethanol 89, 99, 139
Ethanol precipitation 89

F
Farm visit 32, 45, 67, 113, 119
Felids 7
Fowl plague 1, 14, 17, 18
Furin 4, 17

G
Game birds 85, 116
Geese 2, 5, 9, 10, 52, 63, 116, 147
Gel electrophoresis 87, 92, 94, 96, 97, 100, 103, 105, 128
Gene pool 5, 16
Gross lesions 47, 50-52, 56, 59, 61, 63, 65, 69, 70, 117, 119
Guidelines for farm visits 67, 119
Guinea fowl 3, 51, 61, 85, 118
Gumboro disease 70, 115, 121

H
Haemagglutination (rapid HA test) 75, 123
– for chicken sera 83-85
– inhibition test (HI test) 19, 20, 77
– test in petri dishes 76, 123
Haemagglutination test in microtitre plates (micro HA test) 76
Haemagglutinin glycoprotein 1
Horses 6, 15
Host 1, 4, 5, 6, 11, 13, 14, 17, 20, 22, 28, 29, 68, 69, 113, 117, 120, 133
Host range 5, 20, 28
Human health 1, 11, 14, 24, 31

I
Incineration 34, 141, 142, 149, 173
Individual protection device (IPD) 67
Infected premises 23, 32, 137, 159
Infectious bursal disease (Gumboro disease) 121
– laryngotracheitis 69, 70
Intracerebral pathogenicity index (ICPI) 22, 124, 125
Intravenous pathogenicity index (IVPI) 29, 80
IZSVe protocol 89, 101, 108, 110, 117, 127, 129

J
Japanese quails 51, 61, 62

L
Laboratory veterinarian 153
Lithium chloride (LiCl) precipitation 89
Live-bird markets 8

M
Marine mammals 6
Mean death time 124
Microhematocrit tube method 83
Mink 7, 14, 16
Mixing vessel 6
Mustelids 7

N
Necropsy 35, 37-39, 41, 43, 68, 173
Neuraminidase inhibition test (NI) 1, 78, 79, 81, 175
Newcastle disease
– characterisation 22, 30, 123, 124
– clinical signs 20, 24, 28, 29, 32, 33, 113-121
– clinical traits 113
– conventional diagnosis 123
– conventional methods 124
– differential diagnosis 113, 121
– ecology 19
– epidemiological investigation form 33, 155
– epidemiology 19, 20, 28, 32, 139
– farm visit 32, 113, 119
– guidelines for shipping 177
– molecular diagnosis 20, 127
– notification 27-29, 31, 32
– outbreak 19-21, 23, 31, 133, 138

- pathology 113
- serological methods 123
- serological test 19, 124
- virus 24-26, 30, 121-125, 127, 129, 131, 132, 177-179, 181
- virus evaluation of pathogenicity 124

Notifiable avian influenza (NAI) 28, 29

O

Official veterinarian (OV) 32, 153, 164
OIE (World Organisation for Animal Health) 9, 10, 22, 23, 27-30, 33, 73, 87, 108, 123, 151, 152, 177-179
OIE terrestrial code 29, 30
One-Step RT-PCR 87, 101, 127-130
Orthomyxoviridae 1
Ostriches 4, 9, 20, 24, 45, 51, 59, 85, 115, 121, 122, 147

P

Pathogenicity 2, 7, 22, 28, 29, 41, 45, 46, 69, 80, 116, 121, 124, 125
Pathotype 20, 22, 70, 103, 105, 114, 115, 117, 119, 120
PCR (polymerase chain reaction) 87, 91, 170
Period of virus
- introduction (PVI) 33
- spread (PVS) 33

Pet birds 21, 25, 26, 29, 117, 121, 122
Pigeons 2, 20, 21, 24, 25, 116, 121, 146
Pigs 6, 11, 15, 16, 18, 158
Plaque formation 123, 124
Polymerase chain reaction (PCR)
- components 90
- cycling reaction 90
- limitations 92
- mix preparation 95
- organisation of a laboratory 110
- real-time 93

Poultry 1, 2, 4-11, 13, 14, 16, 17, 19-21, 23-26, 28, 29, 31, 33, 35, 41, 45, 52, 55, 67-70, 103, 108, 110, 113, 120-122, 125, 133, 136, 138, 140, 142, 144-150, 163, 170
- farm 67, 68, 120, 135, 138, 148, 163

Preparation of red blood cell suspensions 83
Primary introduction 8, 23
Proteases 1, 4, 22

R

RBCs (red blood cell suspensions) 76, 77, 83, 84, 85
Real-time RT-PCR 87, 92-95, 108, 109
Reassortment 6, 7, 11, 13, 16, 18, 170
Rendering 34, 145, 146, 149, 150
Restriction endonuclease analysis 127, 128
Restrictions 13
Retrotrascription-polymerase chain reaction (RT-PCR)
- end-point protocols 101, 103, 105, 127, 129
- one-step protocols 103, 104, 129
- RT-PCR (Retrotranscription-polymerase chain reaction) 20, 87, 90, 92, 101, 103-105, 107-110, 127-131

RNA
- concentration 89
- extraction 88, 94, 95, 127
- purification 89

RNase inactivation 88

S

Saccharides 5, 6, 11
Safety 68, 78, 134, 137, 139, 142, 146, 152, 170-172
SDS-polyacrilamide gel (SDS-PAGE) 98, 101-106, 127, 129, 130
Seals 6, 7, 14, 16, 17
Serology 80
Sialic acid 5
Species barrier 5
Spectrophotometric method 78, 81, 83
Spread 1, 2, 5-11, 13, 20-28, 30-33, 55, 67, 92, 117, 133, 137-140, 145, 146, 160, 161, 169
Suspicion 20, 31-33, 70, 155, 164
Swabs 23, 39, 42, 43, 68, 73, 88, 94, 95, 102, 111, 137, 153, 177, 178, 181
Swans 6, 10, 116

T

Tracheal / Oropharyngeal swabs 42
Transmission 5-8, 16-18, 22, 23, 28, 53, 138, 145
Trypsin-like enzymes 4, 22
Turkeys 2, 5-9, 14-18, 21, 25, 26, 45-49, 51, 52, 61, 69, 70, 113, 114
Two-Step RT-PCR 87, 105
Types of specimens 177

V

Vaccine 2, 9, 13, 14, 29, 30, 45, 113, 122, 124, 125
Vehicles 9, 32, 33, 117, 133, 134, 138, 145, 146, 162, 163
Viral virulence 21
Virulence 1, 2, 4, 16, 17, 19-23, 25, 26, 29, 30, 70, 80, 87, 115, 119, 120, 124, 127
Virus isolation 35, 39, 43, 73-75, 94, 102, 123, 127, 137

Virus isolation in embryonated eggs 123

W

Whale 7, 15
Wild birds 4-8, 10, 15-17, 19-21, 23, 29, 30, 69, 80, 108, 110, 122, 158
Wild waterfowl 45, 63
Windrow composting 143

Z

Zoonosis 24
Zoonotic potential 27, 28

Printed in December 2008